Journal of Organometallic Chemistry Library 16

Organotin Compounds in Modern Technology

Journal of Organometallic Chemistry Library

1 D. SEYFERTH (Editor)
   NEW APPLICATIONS OF ORGANOMETALLIC REAGENTS IN ORGANIC SYNTHESIS
   Proceedings of a symposium at the American Chemical Society National Meeting held in New
   York City, April 6 — 9th, 1976
2 D. SEYFERTH (coordinating editor), A.G. Davies, E.O. FISCHER, J.F. NORMANT and
   O.A. REUTOV (editors)
   ORGANOMETALLIC CHEMISTRY REVIEWS: ORGANOSILICON REVIEWS
3 D. SEYFERTH (coordinating editor). A.G. DAVIES, E.O. FISCHER, J.F. NORMANT and
   O.A. REUTOV (editors)
   ORGANOMETALLIC CHEMISTRY REVIEWS
4 D. SEYFERTH and R.B. KING (editors)
   ORGANOMETALLIC CHEMISTRY REVIEWS; ANNUAL SURVEYS: SILICON — TIN —
   LEAD
5 D. SEYFERTH (coordinating editor), A.G. DAVIES, E.O. FISCHER, J.F. NORMANT and
   O.A. REUTOV (editors)
   ORGANOMETALLIC CHEMISTRY REVIEWS
6 D. SEYFERTH and R.B. KING (editors)
   ORGANOMETALLIC CHEMISTRY REVIEWS; ANNUAL SURVEYS: SILICON —
   GERMANIUM — TIN — LEAD
7 D. SEYFERTH (coordinating editor), A.G. DAVIES, E.O. FISCHER, J.F. NORMANT and
   O.A. REUTOV (editors)
   ORGANOMETALLIC CHEMISTRY REVIEWS
8 D. SEYFERTH and R.B. KING (editors)
   ORGANOMETALLIC CHEMISTRY REVIEWS; ANNUAL SURVEYS: SILICON —
   GERMANIUM — TIN — LEAD
9 D. SEYFERTH (coordinating editor), A.G. DAVIES, E.O. FISCHER, J.F. NORMANT and
   O.A. REUTOV (editors)
   ORGANOMETALLIC CHEMISTRY REVIEWS; including selected plenary lectures from the
   fifth International Symposium on Organosilicon Chemistry held in Karlsruhe, August 14—18,
   1978
10 D. SEYFERTH and R.B. KING (editors)
   ORGANOMETALLIC CHEMISTRY REVIEWS; ANNUAL SURVEYS: SILICON —
   GERMANIUM — TIN — LEAD
11 D. SEYFERTH and R.B. KING (editors)
   ORGANOMETALLIC CHEMISTRY REVIEWS; ANNUAL SURVEYS: SILICON — TIN —
   LEAD
12 D. SEYFERTH (coordinating editor), A.G. DAVIES, E.O. FISCHER, J.F. NORMANT and
   O.A. REUTOV (editors)
   ORGANOMETALLIC CHEMISTRY REVIEWS; including Plenary Lectures from the 3rd
   International Conference on the Organometallic and Coordination Chemistry of Germanium, Tin
   and Lead (Dortmund, July 21st—25th, 1980)
13 D. SEYFERTH and R.B. KING (editors)
   ORGANOMETALLIC CHEMISTRY REVIEWS; ANNUAL SURVEYS: SILICON —
   GERMANIUM — TIN — LEAD
14 R.B. KING and J.P. OLIVER (editors)
   ORGANOMETALLIC CHEMISTRY REVIEWS; ANNUAL SURVEYS: SILICON — LEAD
15 O. STROUF, B. CASENSKY and V. KUBANEK
   SODIUM DIHYDRIDO—BIS(2-METHOXYETHOXO)ALUMINATE (SDMA).
   A Versatile Organometallic Hydride

Journal of Organometallic Chemistry Library 16

Organotin Compounds in Modern Technology

COLIN J. EVANS AND STEPHEN KARPEL

Development Department, International Tin Research Institute, Fraser Road, Perivale, Greenford, Middlesex UB6 7AQ, England

ELSEVIER

Amsterdam — Oxford — New York — Tokyo 1985

ELSEVIER SCIENCE PUBLISHERS B.V.
Molenwerf 1
P.O. Box 211, 1000 AE Amsterdam, The Netherlands

Distributors for the United States and Canada:

ELSEVIER SCIENCE PUBLISHING COMPANY INC.
52 Vanderbilt Avenue
New York, N.Y. 10017

ISBN 0-444-42422-9 (Vol. 16)
ISBN 0-444-41445-2 (Series)

Printed in The Netherlands

CONTENTS

Preface

VIII

PREFACE

In just a few decades organotin compounds have developed from being laboratory curiosities to large scale industrial chemicals. One of the first uses of
organotins, and still the most important in terms of consumption, was as stabilisers for PVC. Following research into the biocidal properties of certain compounds, new avenues of application were opened, in the fields of wood preservation, crop protection and marine antifouling coatings. These uses for the most
part arose out of intensive research, some of it sponsored by the International
Tin Research Council and later continued in the Council's research laboratories.
Today all over the world, there is a large body of research being conducted into
organotin properties and many of the findings will no doubt eventually lead to
further uses; a good example exists in the case of the mono-organotin compounds,
which have low mammalian toxicity and are showing potential as water repellents
for fabrics and building materials.

The interest in organotin chemicals is reflected in the enormous number of
research papers which are published each year on this subject. There are also a
number of excellent books covering the properties and general applications.
When it comes to examining the industrial use of these compounds in more detail,
a problem arises; this is due to the complexity of the technologies in which organotins are often used. To follow the use of organotin compounds in particular
applications, one has to consult books on PVC technology, or on polyurethane
polymers or wood preservation and hope to find the appropriate sections referring to the tin compounds used. When it comes to more wide-ranging topics such
as agricultural chemicals or medicine it becomes even more difficult to find
suitable coverage of organotin compounds.

It was in view of this, that the authors undertook the present book, in which
the organotin compounds are firmly set in the context of their particular industrial usage. Basic chemistry is kept to a minimum and can be found in greater
detail in other texts. Properties are only described inasmuch as they are relevant to a particular technology. What is attempted is to present an overall
view of each field of application, giving as much of the specialised subject as
is necessary for an understanding of the manner in which the organotin benefits
the process or end use. Thus there are chapters on PVC stabilisation, polyurethane technology, wood preservation, antifouling coatings, agricultural chemicals, other biocidal uses and medical applications. Whilst not claiming by any
means to be complete treatises on these subjects, it is hoped that enough information is provided to give a grasp of the subject and to allow access to more

specific sources when necessary. In this respect, the extensive references are indispensable. A basic introduction to organotin compounds, with brief details of manufacture is provided in the first chapter, in order to put their subsequent use in perspective. One chapter is devoted to the mono-organotins as a newly emerging group of industrial chemicals.

A very important aspect of organotins, as with many other organometallic compounds, is their impact on the environment. In this respect, organotin compounds have an advantage in that they are ultimately degraded to harmless, inorganic forms of tin. The entry into, and subsequent fate of, organotins in the environment are considered in the final chapter, which also contains information on the toxicology of organotin compounds and some toxicity data. Illustrations have been chosen to amplify the text wherever possible.

The authors have for a number of years been engaged in developing the uses of tin chemicals in industry, particularly the organotins and over that period have visited many companies and research organisations, and also held discussions at their Institute. Much of the information has been based on these contacts, and the friendly co-operation of many firms and other organisations who use organotin chemicals in some way is gladly acknowledged here, whether for allowing facilities for visits and photography or for providing photographs or technical assistance.

In particular the authors would like to thank the following:
Dr. A. O. Christie, International Paint plc., Mr. I. D. K. Dring, British Sanitized Ltd., Mr. R. M. Leach, Osmose, Inc., U.S.A., Mr. P. G. Salmon, M and T Chemicals Ltd. The help of colleagues at the International Tin Research Institute, has also been indispensable in producing this book, notably members of Chemistry Division; also Christine Wurm, who prepared many of the drawings, Mr. K. J. Edwards and Sue Miller of the Photographic Studio, who were responsible for many of the photographs in the book.

The authors are grateful to the Director of the International Tin Research Institute and the International Tin Research Council for permission to prepare the book.

Photographic illustrations are the copyright of the International Tin Research Institute, unless otherwise credited by the prefix "Photograph courtesy".

July, 1984

Colin J. Evans
Stephen Karpel

Chapter 1

INTRODUCTION

Organotin compounds can be defined as compounds in which at least one direct tin-carbon bond exists. The great majority of organotin compounds have tin in the IV+ oxidation state although a few are known which have tin in the II+ oxidation state. Tin-carbon bonds are in general weaker and more polar than those formed in organic compounds of carbon, silicon or germanium and organic groups attached to tin are more readily removed. This higher reactivity does not, however, imply any instability of organotin compounds under ordinary conditions.

Four series of organotin compounds are known; those with one tin-carbon bond are known as mono-organotins, those with two tin-carbon bonds are diorganotins whilst triorganotins and tetra-organotins have, respectively, three and four tin-carbon bonds. The properties of these classes of compound differ; thus whilst monoorganotins have low toxicity the triorganotins are characterised by high biocidal activity and diorganotins find use in the plastics industry as stabilisers and catalysts. The tetra-organotins have not found much commercial use to date.

The tin-carbon bond is stable to water and to atmospheric oxygen at normal temperatures and is quite stable to heat (many organotins can be distilled under reduced pressure with little decomposition). Strong acids, halogens and other electrophilic reagents readily cleave the tin-carbon bond. Tin forms predominantly covalent bonds to other elements but these bonds exhibit a high degree of ionic character, with tin usually acting as the electropositive member. Triorganotin hydroxides behave not as alcohols but more like inorganic bases, although strong bases remove the proton in certain triorganotin hydroxides since tin is amphoteric. Thus, bis(triorganotin) oxides, $(R_3Sn)_2O$, are strong bases and react with both inorganic and organic acids, forming normal salt-like but non-conducting and water-insoluble compounds. Tin doubly bonded to oxygen does not exist and diorganotin oxides, R_2SnO, are polymers, usually highly cross-linked via intermolecular tin-oxygen bonds. Unlike the halocarbons, organotin halides are reactive compounds and because of their ionic character, readily enter into metathetical substitution reactions resembling the inorganic tin halides. Unlike carbon, tin shows much less tendency to form chains of tin atoms bonded to each other. Although tin-tin bonded compounds are known (for example hexa-organoditins) the tin-tin bond is easily cleaved by oxygen, halogens and acids.

1.1   HISTORICAL DEVELOPMENT

The first organic compound of tin was prepared in 1852 by Lowig, but credit for the first comprehensive study of organotins belongs to Sir Edward Frankland (1825-1899). In 1853 he prepared "di-iodo diethylstannane" (diethyltin di-iodide) and in 1859 "tetraethylstannane" (tetra-ethyltin); other compounds followed. Despite this early discovery, organotin compounds remained little more than laboratory curiosities for nearly a century.

The first commercially significant property of organotins to become recognised, was the ability of diorganotins to inhibit the degradation, under the influence of heat and UV light, of polyvinyl chloride (PVC). Diorganotin compounds were introduced as PVC stabilisers in the U.S.A. in the 1940s, in the U.K. in 1951 and in the rest of Europe and Japan in the mid-1950s. Developments in this field have been reviewed by Verity-Smith [1]. World production of organotin compounds was stated to be running at 1000-2000 tonnes by 1957, mainly for dibutyltin stabilisers for PVC. More sophisticated compounds were soon developed, in particular, thiotin compounds were shown to be extremely efficient heat stabilisers for PVC. Early in 1955, non-toxic octyltin stabilisers for PVC were discovered and following extensive testing, Government approval was obtained in the U.S.A. and elsewhere for the use of tin-stabilised PVC in food-contact applications. PVC bottles for squash and for food products were marketed in the late 1960s and represented a new outlet for octyltin compounds.

The biocidal uses of organotins stemmed from the systematic study of these compounds sponsored at the Institute for Organic Chemistry, TNO, Utrecht, by the International Tin Research Council in 1950. The research team at TNO under the leadership of Professor G.J.M. van der Kerk made important contributions to the study of the organometallic chemistry of tin, synthesising new compounds and establishing their characteristics [2-10]. In particular the powerful biocidal properties of the trialkyltin and triaryltin derivatives were established. In 1957 the German company Hoechst developed an important agricultural fungicide tradenamed "Brestan", based on triphenyltin acetate. This was particularly effective against the damaging potato blight phytophthora infestans. Subsequently in 1961 Philips-Duphar marketed a crop protectant based on triphenyltin hydroxide (which had not been covered in the earlier triphenyltin patent filed by Hoechst). By the mid-1960s world annual production of organotins had risen to around 10,000 tonnes.

Fig. 1.1. The M & T Chemicals b.v. plant at Vlissingen-Oost, The Netherlands.
Photograph courtesy: M & T Chemicals b.v.

In 1959 the seeds of another important use of organotin compounds were sewn,
when the Osmose Wood Preserving Company of America began active investigation
into the possibilities of organotins for wood preservation. As a result of
these studies, the company marketed a commercial organotin-based wood preservative in 1960. This stimulated major developments in Europe and organotins became firmly established in the field of wood preservation over the next decade
in many parts of the world. In the early 1960s organotin compounds began to
find use in antifouling paint systems for ships in view of their biological
activity against a wide spectrum of fouling organisms. By 1970 International
Red Hand Marine Coatings had applied organotin-based coatings to some 200 vessels, representing 5 m gross tonnes of shipping. The subsequent development of
more effective coatings with long-term activity has led to a marked increase in
the use of organotins for antifouling systems and today this market is a growth
area.

In the late 1960s the miticidal properties of tricyclohexyltin hydroxide were
discovered and a commercial product based on this compound has been found
extremely effective in protecting a wide range of crops and ornamentals. In the
following years a number of other active organotin miticides were developed,
namely bis(trineophyltin) oxide and 1-tricyclohexylstannyl - 1,2,4-triazole.

Growth of organotin usage has been considerable over the last few decades as Fig. 1.2 shows. A current estimate of consumption is over 30,000 tonnes per year. Table 1.1 lists principal uses of organotin compounds; these are discussed in depth in the remaining chapters of this book. It should be noted that three general areas of application account between them for over 95% of the total usage. These areas are: PVC stabilisers, agricultural uses and industrial biocides (in paints, timber, etc.).

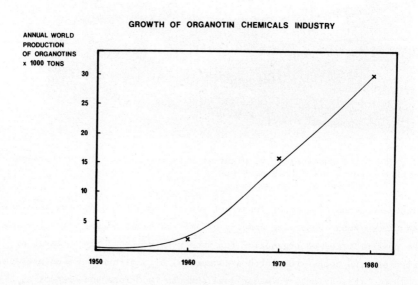

Fig. 1.2. Growth of consumption of organotin compounds.

1.2   METHODS OF MANUFACTURE

Only a brief summary is given here of the manufacturing techniques employed for organotin compounds; more detailed accounts of the chemistry [11] and the process technology [12] have been published. The principal routes are summarised in Fig. 1.3.

1.2.1   Grignard method

A very large proportion of the industrially important organotin compounds is prepared by the Grignard method:

$$SnCl_4 + 4RMgCl \rightarrow R_4Sn + 4MgCl_2$$

The route provides certain problems in control, since mixed solvent systems are needed and large volumes are involved. However, the process is very

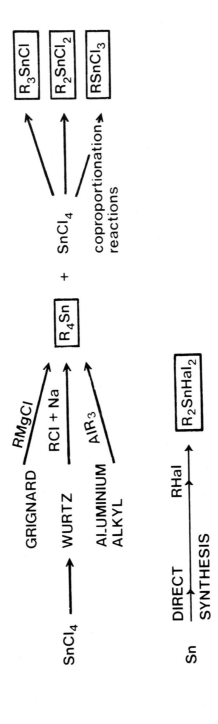

Fig. 1.3. Principal synthesis routes for producing compounds with direct tin-carbon bonds (organotins).

flexible and produces high yields; it is the only industrial method currently feasible for producing phenyltin compounds and it is also employed to produce propyl-, butyl- and octyltin compounds.

1.2.2 Wurtz synthesis

In the Wurtz synthesis, sodium is used in place of magnesium:

$$8Na + 4RCl + SnCl_4 \longrightarrow R_4Sn + 8NaCl$$

Again, large quantities of solvent are required and competing side reactions are a problem in achieving a satisfactory yield. Better results are obtained when the Wurtz synthesis is carried out with alkyltin chlorides rather than with tin tetrachloride. A starting amount of dibutyltin dichloride is first converted by the Wurtz synthesis to tetrabutyltin which then yields twice the amount of dibutyltin dichloride with tin tetrachloride. Half of the dibutyltin compound is then used to produce more tetrabutyltin for further reaction. To date the only company known to be making commercial use of this process is in the German Democratic Republic [13].

1.2.3 Aluminium alkyl route

The aluminium alkyl route to organotins has certain advantages over the other two processes, namely only a small reaction space is required, the process can be operated continuously and no solvents are needed. The method is used in F.R. Germany to produce tetra-alkyltins, particularly tetra-octyltins. Complete transfer of all the alkyl groups bound to the aluminium requires the presence of a complexing agent such as sodium chloride, ethers or tertiary amines. The alkylation to tetra-alkyltin then proceeds smoothly and completely:

$$4R_3Al + 3SnCl_4 + 4R_2'O \longrightarrow 3R_4Sn + 4AlCl_3 \cdot R_2'O$$
$$\text{(complexing} \qquad\qquad \text{(complex)}$$
$$\text{agent)}$$

The aluminium alkyl method is as versatile as the Grignard reaction with regard to the nature of the alkyl groups concerned.

1.2.4 Direct synthesis

Direct synthesis involves reaction between metallic tin and alkyl halide:

$$Sn + 2RHal \longrightarrow R_2SnHal_2$$

The method has certain limitations, for example only dialkyltin species can be produced and only iodides give a suitably high reaction rate, so that the process is only industrially viable when iodine can be economically recovered. The higher reaction temperatures also cause increased dehydrohalogenation as a side reaction, particularly with the alkyl groups of longer chain length. Catalysts have now been developed which enable dibutyltin and dioctyltin dihalides as well as dimethyltin dihalides to be produced by this method.

1.2.5 Coproportionation

To get from the tetra-organotin to the tri-, di-, and mono-organotin compounds, a process known as coproportionation is used. The tetra-organotin is reacted with tin tetrachloride, whereupon redistribution occurs:

$$3R_4Sn + SnCl_4 \longrightarrow 4R_3SnCl$$

$$R_4Sn + SnCl_4 \longrightarrow 2R_2SnCl_2$$

Mixtures of tetra-alkyltin and trialkyltin chloride, as obtained in the Wurtz synthesis, can also be treated in this way. It is sufficient to heat the components together for several hours at 200°C and no solvents are required. In the case of alkyltin trichlorides, $RSnCl_3$, a partial coproportionation of tetra-alkyltin and tin tetrachloride takes place:

$$R_4Sn + 2SnCl_4 \longrightarrow 2RSnCl_3 + R_2SnCl_2$$

the reaction products being subsequently separated by vacuum distillation.

1.3 RANGE OF APPLICATIONS

Organotin compounds today find use in a surprisingly wide range of products; a selection is shown in Fig. 1.4.

Generally speaking, about two thirds of the consumption of organotin compounds takes place in the plastics industry, where mono- and diorganotin compounds find use as stabilisers for PVC and as catalysts in the production of polyurethane foams and room temperature vulcanising (RTV) silicones. Most of the remainder is accounted for by industrial biocides; these find use in wood preservation, crop protection, antifouling paints and in disinfection. Principal manufacturers of organotin compounds are listed in Table 1.2.

Fig. 1.4. Industrial uses of organotin compounds.

TABLE 1.1

Principal industrial uses of organotin compounds

Application	Function	Principal compounds used
PVC stabilisation	Stabilisation against effects of heat and light	dialkyltin di-isooctylthio-glycolate (alkyl = methyl, butyl, octyl, 2-butoxy-carbonylethyl) dialkyltin maleate (alkyl = methyl, butyl, octyl) mono-alkyltin tri-isooctylthio-glycolate (alkyl = methyl, butyl, octyl, 2-butoxy-carbonylethyl
Polyurethane foams RTV silicones	Homogeneous catalysis	dibutyltin diacetate dibutyltin di-octoate dibutyltin dilaurate
Esterification	Homogeneous catalysis	butanestannonic acid dibutyltin diacetate dibutyltin oxide
Glass treatment	Precursor for tin(IV) oxide films on glass	dimethyltin dichloride butyltin trichloride methyltin trichloride
Poultry management	Anthelminthic	dibutyltin dilaurate

Wood preservation	Fungicide	bis(tributyltin) oxide tributyltin naphthenate tributyltin phosphate
Agricultural chemicals	Fungicide Insecticide Miticide Antifeedant	triphenyltin acetate triphenyltin hydroxide tricyclohexyltin hydroxide fenbutatin oxide 1-tricyclohexylstannyl - 1,2,4-triazole
Antifouling paints	Biocide	triphenyltin chloride triphenyltin fluoride bis(tributyltin) oxide tributyltin fluoride tributyltin chloride tributyltin acrylate polymers
Materials protection (stone, leather, paper, etc.)	Fungicide Algicide Bactericide	bis(tributyltin) oxide tributyltin benzoate
Moth proofing of textiles	Insecticide Antifeedant	triphenyltin chloride triphenyltin hydroxide
Disinfection	Bacteriostat	tributyltin benzoate

The mono-organotin compounds, apart from their use in PVC stabilisation, have found little industrial use as yet, but their potential for use as water repellants and fire retardants for fabrics, and as ore flotation agents is being explored. Further details are given in Chapter 9.

Although tetra-organotins have found some use as stabilisers for transformer oils, the principal use of tetra-organotins is as intermediates for other organotin compounds. They have also been suggested as esterification catalysts and in combination with aluminium chloride and certain transition metal chlorides as a catalyst for the low-pressure polymerisation of olefines. The organotin acts as an in situ alkyl acting agent for $AlCl_3$ and avoids the handling of extremely reactive organo-aluminium compounds. However there are no commercial details of their use, and tetra-organotin compounds are not discussed further in this book.

TABLE 1.2

Principal manufacturers of organotin compounds *

EUROPE

ACIMA Chemical Industries Ltd., Buchs/SG, Switzerland
Akzo Chemie B.V., Amersfoort, The Netherlands
Akzo Chemie (U.K.) Ltd., Littleborough, U.K.
Albright and Wilson Ltd., Oldbury, U.K.
Ciba Geigy Marienberg GmbH, Bensheim, F.R. Germany
Commer SRL, Lodi, Italy
Diamond Shamrock Europe Ltd., Manchester, U.K.
Hoechst AG, Burgkirchen, F.R. Germany
Metallgesellschaft AG, Frankfurt/Main, F.R. Germany
M & T Chemicals B.V., Vlissingen, The Netherlands
Polytitane S.A., Vineuil-St.-Firmin, France
Schering AG, Bergkamen, F.R. Germany
Thiokol/Carstab Corp., Maaseik, Belgium
VEB Chemiekombinat, Bitterfeld, Democratic Republic of Germany

JAPAN

Adeka Argus Chemical Co. Ltd., Saitama
Hokko Chemical Industry Co. Ltd., Tokyo
Katsuka Kako K.K., Saitama
Koriyama Kasei Co. Ltd., Tokyo
Nitto Kasei Co. Ltd., Osaka
Sankyo Organic Chemical Co. Ltd., Kanagawa
Tokyo Fine Chemical Co. Ltd., Tokyo
Yoshitomi Pharmaceutical Ind. Ltd., Osaka

U.S.A.

Argus Chemical Corp., Brooklyn, NY
Cardinal Chemical Co., Columbia, SC
Interstab Chemicals Inc., New Brunswick, NJ
M & T Chemicals Inc., Rahway, NJ
Synthetic Products Co., Div. Dart Industries Inc., Cleveland, OH
Tenneco Chemicals, Piscataway, NJ
Thiokol/Carstab Corp., Cincinnati, OH

---

*These are primary producers of compounds containing tin-carbon bonds, as opposed to manufacturers who produce various derivatives starting from an organotin compound as source.

REFERENCES

1   H. Verity-Smith, Development of the Organotin Stabilisers, Int. Tin Res. Inst., Publicn. 302, 27 pp.
2   G.J.M. van der Kerk and J.G.A. Luijten, J. Appl. Chem., 4 (1954) 301-307.
3   G.J.M. van der Kerk and J.G.A. Luijten, ibid., 314-319.
4   G.J.M. van der Kerk and J.G.A. Luijten, ibid., 6 (1956) 49, 56, 93.
5   G.J.M. van der Kerk, J.G. Noltes and J.G.A. Luijten, ibid., 7 (1975) 356-365.
6   G.J.M. van der Kerk, J.G. Noltes and J.G.A. Luijten, ibid., 7 (1957) 366-369.
7   G.J.M. van der Kerk and J.G.A. Luijten, ibid., 7 (1957) 369-374.
8   G.J.M. van der Kerk and J.G. Noltes, ibid., 9 (1959) 106-113, 176-179.
9   J.G.A. Luijten and G.J.M. van der Kerk, ibid., 11 (1961) 35.
10 J.G.A. Luijten and G.J.M. van der Kerk, ibid., 11 (1961) 38.

11   A.G. Davies and P.J. Smith in Sir G. Wilkinson, F.G.A. Stone and E.W. Abel
     (Eds.), Comprehensive Organometallic Chemistry, Pergamon Press, Oxford,
     1982, 519-614.
12   A. Bokranz and H. Plum, Topics Current Chem., 16 (1971) 365-403.
13   U. Thust, Tin and its Uses, 122 (1979) 3-5.

Chapter 2

PVC STABILISATION

2.1   INTRODUCTION

Poly(vinyl chloride) or PVC is one of the best-known and most widely-used of the man-made polymers, and over the last forty years or so has evolved from being a curiosity or mere substitute for other materials to a very important mass-produced commodity in its own right.   In the early 1980s, the consumption of PVC resins in the major industrialised regions of the world includes some 2.5 million tonnes in the U.S.A., 3.5 million tonnes in Western Europe, and 1 million tonnes in Japan.

The key to the success of PVC in so many fields is based on its great versatility, as well as being a relatively low cost material.   Its properties can be modified by incorporating into the polymer a great range of additives, and in fact these additives have a much more important and diverse role in PVC formulation than in any other polymer system.   Thus the basic polymer can be modified to produce either a rigid or plasticised material, and such properties as impact strength, weatherability, flow properties, heat and flame resistance, electrical properties, etc., can all be tailored to meet particular end requirements.   This versatility enables PVC to be processed via all the common techniques for polymers, including calendering, extrusion, injection moulding, blow moulding, vacuum- and press-forming, and thus there are few limits on what can be produced in this material if desired.

The worldwide importance of PVC in many fields has been achieved despite one potentially damaging drawback; like many organic polymers, it is subject to degradation by the action of heat or short-wavelength light.   Consequently, the addition of suitable stabilisers is essential in any particular formulation, since, even if the material is not subjected to extreme conditions in its final form, high temperatures are used (of the order of 200°C) during the processing. Numerous types of stabiliser system have been developed, and organotin compounds form one of the most important groups of these, with the result that PVC stabilisation is responsible for the largest single usage of organotin compounds at the present time, amounting to approximately 20,000 tonnes annually.

The formulation of PVC compound used, and hence the type of stabiliser employed, will vary considerably according to application, and organotins have had

Fig. 2.1. Organotin-stabilised PVC is established in a wide range of outlets.
Photograph courtesy: Akzo Chemie (U.K.) Ltd., Liverpool, U.K.

their main impact in the areas of construction and packaging. The end-uses for tin-stabilised PVC are considered in more detail in section 2.5.

2.1.1   The production of PVC

The route from raw materials to finished articles proceeds basically in two main stages. In the first stage, PVC polymer is produced from the basic feed-stocks in a powder or granular form. This material is subsequently purchased by the processors, who compound it with the various additives required, including heat stabilisers, and convert it into sheet, tubing, etc., for subsequent manu-facture of the finished products, or process the compound directly to finished articles such as injection-moulded items or blow-moulded bottles. The manufact-uring process used for the PVC polymer can have a marked effect on its purity, which in turn affects its stability and thus the quantity and type of stabiliser needed. The processing characteristics and physical/mechanical properties of the polymer are also related to its manufacture, so that it is necessary to choose the right type for a particular application, in addition to formulating the overall compound correctly.

The starting point for PVC is vinyl chloride monomer (VCM) which is produced from the basic feedstocks of ethylene and chlorine. The polymerisation of vinyl chloride is a classical radical reaction, triggered off by an initiator that forms the starting radicals.

Initiation

$$R\cdot + CH_2 = CHCl \longrightarrow RCH_2CHCl\cdot$$

The double bond is opened, and the radical chain grows during propagation.

Propagation

$$RCH_2CHCl\cdot + nCH_2 = CHCl \longrightarrow R(CH_2CHCl)_n CH_2CHCl\cdot$$

Chain termination can occur either by two radicals combining, or by a dispro-portionation reaction that results in one polymer molecule ending in a double bond.

Termination

$$2R(CH_2CHCl)_n CH_2CHCl\cdot \nearrow R(CH_2CHCl)_n CH_2CHCl.CHClCH_2(CH_2CHCl)_n R$$

$$\searrow R(CH_2CHCl)_n CH_2CH_2Cl + R(CH_2CHCl)_n CH = CHCl$$

This is a simple outline of the polymerisation process, but of course in practice the situation can be more complex than this, with various other chain transfer reactions occurring. This can, for example, lead to branching of the polymer chain. A detailed account of the kinetics of vinyl chloride polymerisa-tion is not required here, but can be found elsewhere [1].

There are three principal methods used to carry out this process industrial-ly, namely suspension, emulsion, and mass polymerisation techniques. The first of these is the most common technique used, but all three have been steadily in-creasing with the consumption of PVC; out of a total world nameplate manufactur-ing capacity of 12.5 m tonnes in the mid-1970s, 9.8 tonnes was by the suspension process, 1.7 m tonnes by emulsion, and 1.0 m tonnes by mass polymerisation.

(i)   Suspension polymerisation   In this process, the monomer is dispersed in water in the form of very fine droplets, which are prevented from coagulating by the presence of a protective colloid (usually a water-soluble cellulose or vinyl polymer). A monomer-soluble initiator is used, usually a peroxide, perester, peroxydicarbonate, or azo compound, and the whole mixture is agitated in an autoclave with a water-jacket for temperature control. Because of the slowing down of the reaction at high conversions, it is uneconomic to achieve 100% conversion to polymer, and so the process is stopped at about 70%-90% completion (as determined by the pressure or reaction time). The reaction is highly exothermic, and the kinetics mean that the molecular weight is mainly governed by the polymerisation temperature. However, the rate of polymerisation is controlled by the type of initiator used.

The polymer obtained is in the form of particles of sizes about 10-100 microns; the mixture is stripped of residual vinyl chloride monomer, and the slurry is centrifuged to leave a wet cake of PVC. This is then washed and dried with hot air to leave the final product. Suspension PVC is quite pure, and contains only minimal traces of the additives used. It has good flow properties, and the particle structure may be either glass hard or porous, the latter type being advantageous in producing a dry mix even when a large amount of plasticiser is added. Suspension PVC can be used for most applications, except for pastes, and is suitable for high-transparency products in which organotins are usually employed.

(ii)   <u>Emulsion polymerisation</u>   This was the earliest method used to manufacture PVC, and here the monomer is emulsified with water using an emulsifying agent (for example alkyl sulphonates or fatty alcohol sulphates) in an autoclave. In conventional emulsion polymerisation, the process takes place in the presence of a water-soluble initiator such as ammonium or potassium persulphate. The PVC particles thus obtained are in a very fine dispersion latex, with sizes less than 1 micron; the rate of initiation and the emulsifier concentration can be used to control the particle size. The PVC dispersion is too fine to be centrifuged, and all the water must be removed by evaporation. This means that the polymerisation additives used are retained in the PVC, affecting its properties; the emulsifiers present facilitate processing, but reduce weather-fastness, electrical insulation properties, and heat stability.

An alternative technique of emulsion polymerisation, called the microsuspension process, uses a monomer-soluble initiator, and a continuous polymerisation method has also been developed as an alternative to these batch processes.

Emulsion PVC is used for speciality applications such as thin profile extrusions where particularly good processing characteristics are required; it is also used in paste technology for similar reasons.

(iii)   <u>Mass polymerisation</u>   This method (also called block or bulk polymerisation) depends on the fact that PVC is insoluble in its monomer, and will precip-
itate out from it. The process is carried out in two stages, a pre-polymerisation to produce polymer seed particles, and a post-polymerisation where the final desired molecular weight is achieved, usually employing up to 80% conversion
of monomer. The pre-polymerisation involves high speed agitation, and the speed
of the agitator controls the particle size of the seed, and hence of the final
product. The porosity of the final PVC is controlled by the temperature of the
pre-polymerisation, which can thus be adjusted to produce either plasticised or
rigid extrusion grades. After the pre-polymerisation, the material is fed into
autoclaves and mixed with additional monomer and initiator. The reaction continues, the temperature controlling the final molecular weight, and the PVC powder can be separated from the mixture in a dry state. Mass-polymerised PVC is
quite pure, containing only traces of the initiator used, and it is particularly
suitable for high-transparency products.

In addition to the above processes, PVC can also be obtained by a solution
technique, where a solvent medium is used for either the monomer and polymer
(homogeneous system) or just for the monomer, with the polymer precipitating out
(heterogeneous system). High purity PVC is obtained from solution polymerisation.

All polymerisation processes produce a range of molecular weights in the final product, but the molecular weight distribution in a given PVC is relatively
narrow, due to the polymer structure being fairly simple. Thus there is little
long chain branching (averaging less than one per PVC molecule), and comparatively few short chain branches (between two and five per 1000 carbon atoms, arising from molecular rearrangement during polymerisation) [2]. The average molecular weight of PVC normally falls within the range 40,000-129,000.

2.1.2   <u>Vinyl chloride, PVC, and toxicity</u>
Mention must be made of one aspect that has produced considerable changes in
the PVC industry over the last ten years, namely the potential health hazard
due to the monomer vinyl chloride. After nearly fifty years of commercial production of PVC, it was discovered in the early 1970s that prolonged exposure to
vinyl chloride could induce angiosarcoma of the liver in human beings, a rare
form of cancer. An official report sponsored by major PVC manufacturers an-
nounced its findings to this effect in January, 1974, and shortly after, working committees were set up in the U.K. and many other countries to assess the
problem and recommend the changes needed in manufacture and processing.

The areas where possible exposure to vinyl chloride could occur include:
a) The polymerisation operation, b) Storage and handling of the polymer, c) Hot processing of the polymer, d) Fabrication of finished products, and e) Diffusion out of finished products, especially migration into foodstuffs, etc. The main area of concern is the actual polymerisation plant, where concentration levels in the atmosphere around the plant could reach unacceptably high levels. Major changes in operating procedure, however, have since drastically reduced these levels. For example, the typical average vinyl chloride concentration around a plant was in the region of 300-400 ppm in the period 1960-70, about 150 ppm in mid-1973, and down to 2-5 ppm at the end of 1976 [3]. These improvements have been brought about by various modifications, including the introduction of large computer-controlled autoclaves.

Vinyl chloride can also be entrapped in the final resin, the degree of entrapment depending very much on the type of resin; for example those resins with an open porous structure (as used for plasticised compounds) are easier to purify than those with a hard, dense structure (as required for extrusion). Nevertheless, the residual monomer levels in PVC resins, PVC compound blends, and fabricated articles have all been appreciably reduced since 1974, as Table 2.1 illustrates for PVC bottles produced in the U.K.

TABLE 2.1
Residual monomer levels in U.K.-produced PVC bottles and resins [4].

	Monomer concentration (ppm)		
	Jan. 74	Mar. 75	Dec. 76
Level in polymer at time of use	500-1000	100-250	15-50
Typical level in compound at time of manufacture	50	5	1
Typical level in bottle wall	50	3	< 1

As a result of the action taken by the various sectors of the PVC industry, they have been able to meet the strict requirements laid down by various national health authorities, and it can now be truthfully said that PVC has been one of the most closely scrutinised of all man-made packaging materials, and has consequently emerged from this uncertain period with renewed confidence. Although some markets were badly affected in the mid-1970s, particularly in the foodstuffs sector, PVC has made up most of the lost ground since then and has continued to expand in many markets.

2.2 ORGANOTIN STABILISERS

2.2.1 Historical development

The use of organotin stabilisers for PVC started in the U.S.A. in the 1940s
following development work by a series of workers at Union Carbide. The first
patent (U.S. 2,219,463; G.B. 497,879) was filed in December, 1936 by V. Yngve,
and was published in October, 1940. This covered the use of alkyl, aryl, and
alkyl-aryl derivatives of tin and lead as heat stabilisers, and succeeding patents soon after covered the use of oxide and hydroxide derivatives of the above
compounds. However, these formulations were not commercially successful, and
the first patent for a commercially successful stabiliser was for dibutyltin dilaurate and similar compounds published in March, 1944 [5]. These were utilised
for dryspun vinyl chloride-acetate copolymer fibres.

$$nC_4H_9 \diagdown \quad \diagup OOC.C_{11}H_{23}$$
$$Sn$$
$$nC_4H_9 \diagup \quad \diagdown OOC.C_{11}H_{23}$$ Dibutyltin dilaurate

In the years following, many types of organotin compound were patented, but
few of these became commercially significant. One which did included the derivatives of α,β -unsaturated carboxylic acids [6]. This covered the dialkyltin
maleates, which exist in a polymeric form, and which have subsequently become
one of the principal types of stabiliser.

$$\left[\begin{array}{c} R \\ | \\ -Sn-OOC.CH=CH.COO- \\ | \\ R \end{array} \right]_n$$ Dialkyltin maleate

Other compounds patented included organotin alkoxides and various derivatives
[7, 8], alkyltin oxide addition complexes [9], organotin phosphates [10], and
numerous sulphur-containing compounds where there is no direct tin-sulphur bond.
The last-named group included the dialkyltin salts of o-sulphobenzimide [11],
organotin sulphonamides [12], and dialkyltin salts of mercaptoacids [13].

However, the group of compounds that have subsequently come to dominate the
field of organotin stabilisers are those containing a direct tin-sulphur linkage, and the earliest such patent appears to be filed by the Firestone Company
in 1950, although not published until 1956 [14]. This applied to dialkyltin

mercaptides having the general formula:

$$R\diagdown \atop R\diagup Sn \diagup ^{SR'} \atop \diagdown _{SR'}$$

Dialkyltin dimercaptide

It was quickly recognised that this type of compound was a very efficient heat stabiliser, and could be used in low concentrations in the resin. A great number of patents subsequently appeared, covering almost every conceivable orga-notin compound containing an Sn-S bond, and by 1959, there had been over 100 published patents on the use of organotins generally as PVC stabilisers [15].

During the 1950s, commercial production of such stabilisers was underway in the United States at the companies Metal and Thermit Chemicals, Advance Solvents and Chemical Corp., Argus Chemical Corp., and Carlisle Chemical Works, while production started in the U.K. at Pure Chemicals Ltd., in 1952 (with dibutyltin dilaurate) and Albright and Wilson Ltd., in 1957. In 1955, production began in Japan, and also in West Germany at Deutsche Advance of Marienberg; many other countries soon followed.

The use of organotins in rigid extruded PVC pipes was established in the United States in 1953, and this sector has remained an important market, altho-ugh one peculiar to this country; in Europe, PVC pipe is mainly lead-stabilised.

Much PVC is now utilised in the packaging of food and drink, and hence it was an important breakthrough to discover that certain organotin compounds, while being effective heat stabilisers, were also of very low mammalian toxicity and thus could be incorporated in food-contact material. The International Tin Research Institute sponsored research at TNO, Utrecht, during the 1950s into the properties and possible applications of organotins, which resulted in the dis-covery of the biocidal properties of triorganotin derivatives, and the partic-ularly low toxicity of dioctyltin derivatives [16]. The latter compounds were thus investigated as possible additives in food-grade PVC, and the use of cer-tain dioctyltin compounds was approved by the B.G.A. (The German Federal Health Authority) in 1961, for unplasticised PVC intended for non-fatty foodstuffs. The U.S. Food and Drugs Administration (FDA) also subsequently (1968) approved the use of two di-n-octyltins, namely (a) the cis-butene-dioate (or maleate) polymer and (b) the S, S'-bis(isooctylmercaptoethanoate), the latter being a liquid. Alternative chemical names for (b) are the diisooctylmercaptoacetate or diisooctylthioglycolate of di-n-octyltin.

$$\left[\begin{array}{c} nC_8H_{17} \\ | \\ -Sn-OOC.CH=CH.COO- \\ | \\ nC_8H_{17} \end{array} \right]_n \qquad \text{(a)}$$

$$\begin{array}{c} nC_8H_{17} \qquad\qquad SCH_2COO.iC_8H_{17} \\ \diagdown\qquad\diagup \\ Sn \\ \diagup\qquad\diagdown \\ nC_8H_{17} \qquad\qquad SCH_2COO.iC_8H_{17} \end{array} \qquad \text{(b)}$$

Many other countries around the world have subsequently approved these stabilisers for food-contact PVC.

Progress was also made in the early 1960s in the manufacture of the stabilisers, and this was an important factor in maintaining their cost-competitiveness with other systems, since the basic materials involved (particularly the tin) are relatively costly. The manufacturing methods used for organotins are described in Chapter 1, and in most cases the key stage for the economic production of the stabiliser is the initial preparation of the tetraalkyltin, since this compound serves as the starting point for the manufacture of most other organotin compounds. The original process used for tetraalkyltins is the Grignard method using magnesium, but a modified Wurtz process using sodium was introduced in 1960 at VEB Chemiewerk Greiz-Dölau in the German Democratic Republic, and is used here to produce butyltin and octyltin stabilisers [17]. The aluminium-alkyl synthesis was introduced by Schering AG of Bergkamen, German Federal Republic, in 1963. Some large investments in the production of tetraalkyltins by this method have been made recently, for example a new plant commissioned by M and T Chemicals at Axis, Alabama, which is intended to have an initial production of up to 7000 tonnes per annum of tetrabutyltin.

In addition to the above three methods for generating tetraalkyltins, methods were devised in the early 1960s for producing dialkyltin dihalides by a "direct synthesis"; here, tin metal is reacted directly with an alkyl bromide or iodide in the presence of a special catalyst. Processes were developed along these lines by the Yoshitomi Drug Manufacturing Company in Japan, Deutsche Advance Produktion in F.R. Germany, and Pure Chemicals Ltd., in the

U.K. [18]. In 1970, Cincinnati Milacron Chemicals started commercial production of a new type of organotin stabiliser by direct synthesis, namely the methyltins. Whereas most stabilisers up to that point had been either based on butyltin or octyltin, it was found that the methyltins also had extremely good stabilising action and (since they contained a relatively higher proportion of tin) could be used at very low concentrations in the PVC formulation. In addition, the more polar character of the methyl groups compared to butyl or octyl is claimed to lead to better compatibility with the PVC polymer (which is itself rather polar due to the chlorine atoms), and hence promote better clarity and processing characteristics. The methyltins have secured a substantial market in the U.S.A. for rigid pipe extrusions (including potable water piping), and the mono- and dimethyltin isooctylthioglycolates have been approved for food-contact PVC in several European countries, including the German Federal Republic and the U.K.

The latest major development in both the manufacture and the stabiliser compound itself has been the introduction of the "Estertins" by AKZO Chemie in 1976. These have the general formula R_2SnX_2 and $RSnX_3$, where R here is an ester group instead of the usual alkyl group, and X may be a typical anionic group such as thioglycolic acid ester. They are prepared by the in situ reaction of tin metal, hydrogen chloride, and an α, β-unsaturated carboxyl compound, the overall reaction being represented by:

$$2HCl + Sn + 2CH_2 = CHCOOR \longrightarrow Cl_2Sn(CH_2CH_2COOR)_2$$

The chloride intermediate produced can then be converted to the desired stabiliser compound. The amount of monoestertin produced can be controlled by the reaction conditions, but none of the toxic triestertin is formed in the above process. Thus the synthesis is non-hazardous compared to conventional routes, and proceeds from basic starting materials under mild conditions without the aid of a catalyst. The estertins have been found to be excellent heat stabilisers, and are also claimed to have lower odour, lower migration tendency, and better light-stabilising properties than alkyltin mercaptides. They are used in various markets, particularly for rigid extrusions in the U.S.A., and have also been approved for food-contact use in numerous European countries.

Looking at the development of organotin stabilisers since their inception, there has been a continuing trend towards a lower total tin content in the PVC formulation, spurred by the fact that the cost of tin has risen from £800 per tonne in 1957 to about £8,500 per tonne in 1983. Thus there has been a trend to lower loading levels of stabiliser in the resin. The loadings are now commonly between 1 and 2 parts per hundred of resin (phr), but values less than 1 phr are used, and as low as 0.25 phr has been employed in twin-screw extruders. Tin is rarely used now with other metal stabilisers, but further economies have been attempted at various times by the marketing of "mixed-metal" formulations, such as tin-barium, tin-calcium, or tin-antimony, primarily intended for pipe extrusions. These have found some use in the U.S.A.

Lower stabiliser loadings are made possible by, among other things, increased stabiliser efficiency, and one way of achieving this has been to employ a synergistic mixture of compounds, where the net stabiliser efficiency is greater than that of the individual components added together. It was found in the mid-1960s that monoalkyltin compounds, in addition to being stabilisers in their own right, had such an effect when used together with their dialkyltin analogues, and much subsequent research has been done to determine the optimum mono/di ratios for maximum efficiency in particular applications [19]. Today, many commercial organotin mercaptide stabilisers are actually synergistic mixtures, even if the company literature only mentions the main (dialkyltin) component; the proportion of monoalkyltin present may be anything up to 50%. Research carried out by the International Tin Research Institute to clarify the mechanism of

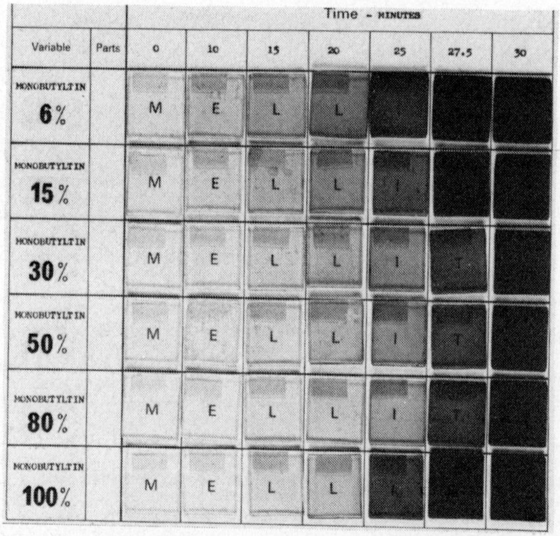

Fig. 2.2. The effect of varying the proportion of monobutyltin mercaptide in a mixture with dibutyltin mercaptide stabiliser is shown in this test series of PVC coupons, using 0.75 phr stabiliser content in an oven ageing test at 190°C. Photograph courtesy: Albright and Wilson Ltd., Warley, U.K.

synergism is described in Chapter 10, on monoorganotin compounds. There has been one monoorganotin that has been developed for use on its own as a stabiliser; this is monobutyltin sesquisulphide $(C_4H_9SnS_{1.5})_4$, which is approved by the B.G.A. for food-contact PVC [20], although it does not appear to enjoy widespread use.

Epoxy compounds that are used as plasticisers in PVC also have a slight synergistic effect with metal-based stabilisers, and improve the long-term heat and light stability of the polymer. For this reason, some stabilisers now marketed comprise an organotin compound blended with about 25% epoxidised soyabean oil. This dilution of the stabiliser also aids the accurate compounding of small quantities in the formulation.

In general, there is a trend today to formulate stabiliser "packages" for the PVC processor in order to make his operations simpler and more reliable. Stabiliser/lubricant packages in a convenient solid form are becoming popular for rigid extrusions, and organotins are among those that are being introduced in such a form.

2.2.2   Compounds employed

A large number of different organotin compounds have been used as heat stabilisers over the years, as described in the last section, but the ones employed today can be grouped into a few main categories. Shown in Table 2.2 are the principal types of compound that are marketed for use as heat stabilisers. Proprietary formulations can sometimes comprise mixtures of compounds (e.g. a maleate plus a mercaptide), in addition to the use of monoorganotin compounds for synergistic reasons.

2.3   THE THERMAL DEGRADATION OF PVC

2.3.1   Degradation processes

PVC is subject to degradation of three main types, namely thermal, photolytic, and biodeterioration. The first of these is our concern in this chapter, and it is the most evident drawback to the processing of the polymer. Degradation becomes visibly apparent by a gradual colour change from white to yellow to brown, and eventually to black if it is allowed to continue. The physical properties of the polymer also markedly deteriorate, becoming increasingly brittle until it starts to disintegrate.

Research into the causes of PVC degradation has been carried out for virtually as long as the material has been produced, and as a result, there is an enormous literature on this subject, which can only be briefly summarised here.

TABLE 2.2

The principal organotin stabilisers

Compound	Formula	
Dialkyltin diisooctylthioglycolate	$\text{R} \diagdown\!\!\!\diagup \text{Sn} \diagdown\!\!\!\diagup \text{R}$ with $SCH_2COO.iC_8H_{17}$ and $SCH_2COO.iC_8H_{17}$	$R = CH_3-$ nC_4H_9- $nC_8H_{17}-$
Dialkyltin maleate	$\left[-Sn(R)(R)-OOC.CH = CH.COO- \right]_n$	$R = nC_4H_9-$ $nC_8H_{17}-$
Dialkyltin maleate ester	$\text{R}\diagdown\!\!\!\diagup Sn \diagdown\!\!\!\diagup\text{R}$ with $OOC.CH = CH.COOR'$ and $OOC.CH = CH.COOR'$	$R = nC_4H_9-$ $nC_8H_{17}-$
Dialkyltin dilaurate	$\text{R}\diagdown\!\!\!\diagup Sn \diagdown\!\!\!\diagup\text{R}$ with $OOC.C_{11}H_{23}$ and $OOC.C_{11}H_{23}$	$R = nC_4H_9-$ $nC_8H_{17}-$
Dibutyltin β-mercaptopropionate	$\text{R}\diagdown\!\!\!\diagup Sn \diagdown\!\!\!\diagup\text{R}$ with $S-CH_2$ / $OOC.CH_2$	$R = nC_4H_9-$
Diestertin diisooctylthioglycolate	$\text{R}\diagdown\!\!\!\diagup Sn \diagdown\!\!\!\diagup\text{R}$ with $SCH_2COO.iC_8H_{17}$ and $SCH_2COO.iC_8H_{17}$	$R = C_4H_9OOC.CH_2CH_2-$

The great volume of work that has been done, and continues to be done, in this area is an indication of the complexity of the subject, and it is fair to say that it is not fully understood yet. However, there is broad agreement on the general principles, even if the evidence for the finer details can vary according to source.

There are two basic processes occurring during degradation, namely loss of hydrogen chloride, and autoxidation, the latter process assisting the former. The removal of hydrogen chloride from the saturated PVC molecule leads to the formation of a double bond, and this site is stated to be able to activate the neighbouring atoms so that further loss of hydrogen chloride occurs. In this way the molecule "unzips" with progressive loss of hydrogen chloride, to leave a conjugated double-bond system:

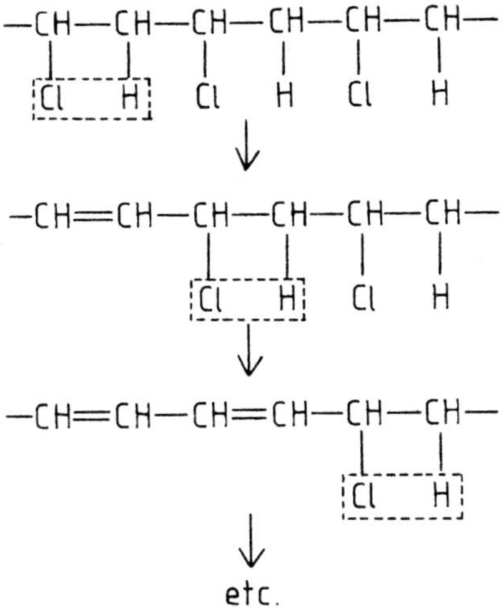

$$-CH-CH-CH-CH-CH-CH-$$
$$\quad Cl \quad H \quad Cl \quad H \quad Cl \quad H$$

$\downarrow$

$$-CH=CH-CH-CH-CH-CH-$$
$$\quad\quad\quad Cl \quad H \quad Cl \quad H$$

$\downarrow$

$$-CH=CH-CH=CH-CH-CH-$$
$$\quad\quad\quad\quad\quad\quad Cl \quad H$$

$\downarrow$

etc.

As the degree of conjugation increases, so the light absorption peak of the molecule shifts from the ultraviolet into the visible region, and the molecule becomes more highly coloured. However, polyene sequences are not the only cause of discolouration, and in fact the marked change in colour that occurs with relatively slight decomposition can not be entirely ascribed to this. It has been shown that the higher polyenes are basic in character and are able to form carbonium ions during the thermolysis of PVC [21]. These carbonium ions become more stable as the polyene length increases, due to the greater delocalisation of the positive charge. They may also incorporate keto structures due to reaction with oxygen (see later).

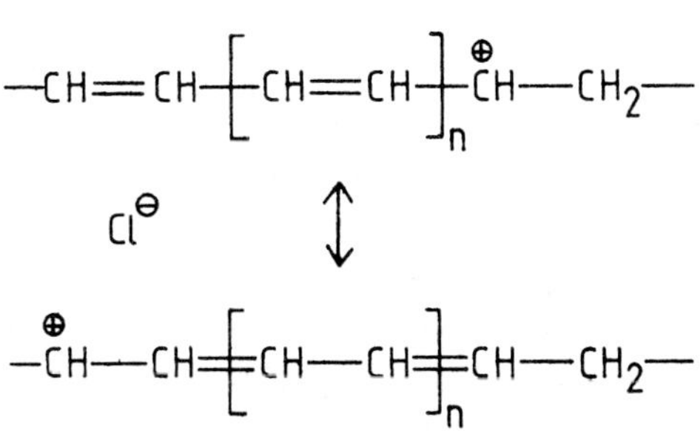

$$-CH=CH-[CH=CH-]_n \overset{\oplus}{C}H-CH_2-$$

$$Cl^{\ominus} \qquad \updownarrow$$

$$-\overset{\oplus}{C}H-CH=[CH-CH=]_n CH-CH_2-$$

The onium salts are strongly coloured, and thus contribute markedly to the darkening of PVC.

2.3.2 The mechanism of dehydrochlorination

A major question is the reason why initial loss of hydrogen chloride should occur at a particular site in the chain, which then subsequently triggers off the "zipper" reaction. It is deduced that commercial PVC must contain a number of "active sites" that are conducive to dehydrochlorination, since a perfectly regular, saturated PVC molecule should be fairly stable against heat. However, commercial PVC starts to decompose in the temperature range 90°-130°C, while analogous low molecular weight model compounds do not start to decompose until 300°-330°C [22].

There are various candidates for possible active sites in the polymer which are discussed in the literature, including the following:

(i)   Chain ends. As shown earlier, a termination by disproportionation will result in one chain having an unsaturated end group. The effect of such groups on the thermal stability of PVC has been extensively studied, but the results are not conclusive. It appears that these end-groups have less effect on the thermal stability than unsaturation within the chain itself. Initiator residues will also appear on some chain ends, but on the whole, the nature of such groups does not appear to be a dominant factor in the thermal stability of the polymer.

(ii)   Branching. Chain transfer reactions during polymerisation can cause branching, with the possible creation of tertiary chlorine atoms (i.e. chlorine bonded to a tertiary carbon atom) at the branch point. These are less thermally stable than primary chlorine atoms, and thus may constitute an active site. However, the existence of such tertiary chlorine atoms is again not conclusive.

(iii)   Unsaturated sites. The presence of a double bond with an adjacent allylic chlorine atom has been shown to be a source of thermal instability, leading to elimination of hydrogen chloride and the start of the zipper reaction. It has been found that up to fifteen randomly distributed double bonds may be present in a PVC molecule per 1000 carbon atoms [23], but the origin of these has not been ascertained; they could arise either during polymerisation or subsequent processing.

(iv)   Oxygenated sites. It is a well established fact that the dehydrochlorination of PVC occurs more rapidly in the presence of oxygen [24].

This has been attributed to various oxidation processes. One such reaction is said to form hydroperoxides as initial active sites, which are subsequently degraded by heat to form olefinic or keto structures that subsequently trigger off the zipper reaction [25]. It has also been stated that the keto-allyl chloride group:

$$\underset{\overset{\displaystyle \|}{\displaystyle -C}}{\overset{\displaystyle O}{}}-CH=CH-\underset{\overset{\displaystyle |}{\displaystyle CH}}{\overset{\displaystyle Cl}{}}-$$

is actually the major cause of thermal dehydrochlorination of PVC [26]. Other autoxidation reactions that affect PVC are described shortly.

The question as to whether the eliminated hydrogen chloride catalyses further dehydrochlorination of the polymer has been argued for many years, and today it is generally accepted that this autocatalytic effect does occur; the process is particularly strong in the presence of oxygen, but also occurs in an inert atmosphere [27].

The mechanism of hydrogen chloride elimination has also been a contentious issue between either free-radical or ionic processes, but it is generally accepted now that the latter is the predominant one in an inert atmosphere. The catalytic effects of hydrogen chloride and other acids, and the acceleration of the reaction in polar solvents, indicate that the elimination probably proceeds via an ionic process and a polar transition state. However, free radical mechanisms may also play a part, particularly if peroxides are formed during degradation in oxygen.

2.3.3 Autoxidation

The thermal degradation of PVC in air involves various oxidation processes in addition to dehydrochlorination, although the two may be closely linked in practice. Three distinct types of oxidation reaction are stated to occur.

In the first, already mentioned, oxidation of the intact polymer chain forms hydroperoxides which subsequently break down and start the zipper reaction [25].

A second type of autoxidation is stated to occur when long polyene sequences have been formed by prior degradation [28]. A "diene synthesis" produces a cyclic peroxide, which can subsequently react further under the effects of heat and evolved hydrogen chloride.

$$-CH=CH-CH=CH- \quad \longrightarrow \quad \begin{matrix} & CH=CH \\ -CH & \quad & CH- \\ & O-O \end{matrix}$$
$$+O_2$$

A third type of autoxidation also occurs during heat degradation, whereby $-CH_2$ or $-CHCl$ groups adjacent to double bonds are vulnerable to attack by oxygen, forming initial hydroperoxides that break down to keto-polyene type structures.

$$\overset{O}{\underset{\|}{-C}} \left[CH=CH \right]_n CH=CH- \qquad \text{Keto-polyene}$$

$$\downarrow +HCl$$

$$\overset{OH}{\underset{\oplus}{-C}} \left[CH=CH \right]_n CH=CH- \qquad \text{Carbonium salt}$$

$$Cl^{\ominus}$$

These keto-polyenes are also able to react further to produce carbonium salts (stabilised by mesomerism), which are very dark in colour even when the polyene sequence is relatively short. It is thought that this type of compound is primarily responsible for the discolouration of heat-aged PVC, although very advanced degradation will produce a more complex chemistry, eventually resulting in elemental carbon.

Thus it will be appreciated that an ideal heat stabiliser for PVC should be able to fulfil several functions within the polymer compound in order to prevent degradation, and it is believed that the effectiveness of organotin compounds in this respect is due to their ability to do this.

2.3.4   The light stability of PVC

Mention should also be made of the degradation of PVC by short wavelength light, since this is an important factor in the formulation for various outdoor applications, and the additives normally termed "heat stabilisers" can also have a marked effect in this area.

Perfectly structured PVC should not undergo any light degradation at all, since it should not absorb any light with wavelengths longer than 300 nm, which is the shortest wavelength radiation that reaches the Earth's surface. However, the presence of irregularities and defects in the polymer enable it to absorb longer wavelengths, and hence initiate degradation processes. As was described in section 2.2.1, these irregularities can include polyene sequences, initiator end-groups, and oxidation structures (e.g. carbonyl groups). The photodegradation of PVC has been studied less extensively than the thermal type, but it appears to be equally complex and is not yet fully understood. As with thermal degradation, the process is accompanied by hydrogen chloride evolution, and the rate is accelerated by the presence of oxygen [29-31]. The mechanism here is stated to be a free-radical one.

In order to counter photodecomposition, PVC that is intended for outdoor use may be compounded with special UV-absorbers (for example, hydroxybenzophenone-type compounds) that absorb strongly in this region and dissipate the energy mainly by molecular vibrations, i.e. heat. The use of efficient heat stabilisers can also aid the light stability, since they minimise the formation of the structure irregularities that absorb UV light (e.g. polyene sequences and oxidation structures). In addition, however, these compounds also have light-stabilising properties of their own, although there appears to be no direct correlation between their efficiencies for thermal and photochemical stabilisation. The organotin mercaptides, for example, are excellent heat stabilisers, but do not give good light stability; the organotin maleates, on the other hand, are less efficient in the former mode but are quite effective for light stability.

The overall light stability of a PVC compound will also depend on the other constituents (such as pigment) and suitable formulations for particular uses must be found by extensive long-period testing outdoors and accelerated xenon-arc lamp tests.

2.3.5   Stabilisation mechanisms

The organotins are very efficient stabilisers, and in particular the mercaptides are the most effective known. This is considered to be due to the fact that they can efficiently fulfil several different functions in the PVC during processing.

(i)   The organotin compound can eliminate the hydrogen chloride that is produced during thermal decomposition:

$$R_2SnX_2 + 2HCl \longrightarrow R_2SnCl_2 + 2HX$$

Since hydrogen chloride has an autocatalytic effect on further dehydrochlorination (especially in the presence of oxygen), this is an important stabilisation mechanism. The dialkyltin chloride produced can volatilise during the processing, which explains why a reduction in tin content of PVC is found to occur at this stage. Some workers consider that the alkyltin chlorides have a further retardation effect on dehydrochlorination by forming protective complexes with allylic chlorine atoms in the PVC, stabilised by mesomerism [32]:

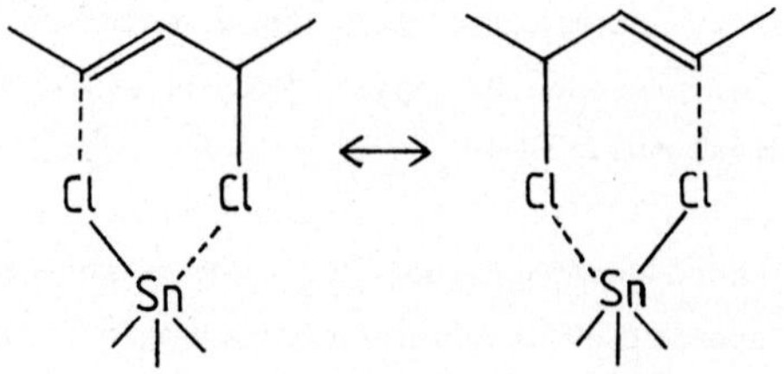

(ii) The exchange of labile chlorine atoms in the PVC (which act as initial sites) with the stabiliser is another important function. A substitution reaction occurs which exchanges the chlorine atom for the "X" group in the stabiliser, e.g. a mercapto group. The mechanism is considered to be [33]:

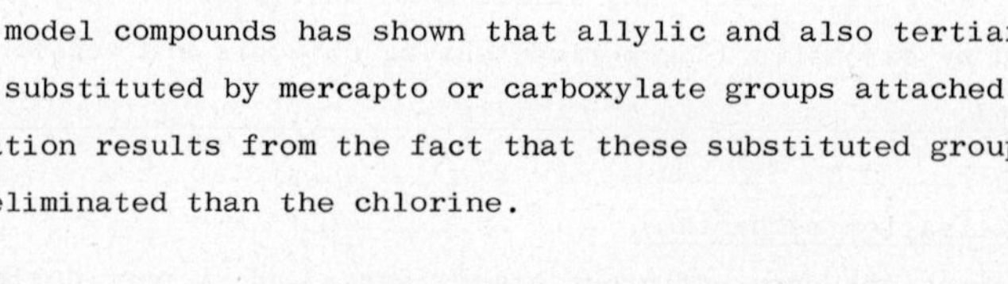

Work with model compounds has shown that allylic and also tertiary chlorine atoms can be substituted by mercapto or carboxylate groups attached to the tin. The stabilisation results from the fact that these substituted groups are much less easily eliminated than the chlorine.

(iii) Organotin mercaptides have an antioxidative action, and since the dehydrochlorination accelerates in the presence of oxygen, this is another stabilising effect. It has been shown that the mercaptides are able to decompose any hydroperoxide groups formed that can act as initiation sites during thermolysis.

(iv)   It is an interesting observation that PVC which has already been darkened by heat treatment can have its colour considerably improved by subsequently treating it with an organotin mercaptide stabiliser [34]. From this it is inferred that the stabiliser is able to react with and destroy the onium salts that are primarily responsible for the colour. In addition, it is thought that the free isooctyl mercaptoacetate ester (which is formed during processing when the organotin mercaptide reacts with hydrogen chloride) is able to shorten any polyene or polyenone sequences by adding to double bonds [35]; in fact, it has been found that a marked colour improvement in heat-degraded PVC can also occur by treatment with isooctyl mercaptoacetate ester [19].

Polyene or polyenone sequences can also be shortened by reaction with maleate derivatives, which leads to the highly favourable Diels-Alder type rearrangement. This may explain why organotin maleates are superior to other carboxylates for preventing discolouration of PVC [34].

2.4   PROCESSING OF PVC

2.4.1   Formulation

PVC is never used as the pure polymer, but is always compounded with other materials in order to achieve the desired processing characteristics and the desired final properties. In addition to the PVC polymer and heat stabiliser, other common compounds employed may include plasticisers, lubricants, pigments, and fillers.

Plasticisers are able to substantially modify the mechanical properties of PVC. By penetrating between the polymer chains, the plasticiser molecules enhance their relative mobility and lower the softening point. Plasticisers are polar compounds, the most common being phthalate esters, and additional specific properties can be imparted by using particular families of plasticisers; these may include low-temperature properties, flame retardancy, extraction resistance, stain resistance, etc. However, plasticisers are not an essential component, and a large part of PVC processing is done with non-plasticised or rigid formulations. Tin-based stabilisers are mainly used in non-plasticised compounds due to the greater efficiency required during processing, but they are also used in some types of plasticised PVC, including paste formulations.

Lubricants are an important constituent to aid the processing of PVC, and are essential for rigid formulations. They are sometimes divided between internal lubricants (to reduce friction between the actual polymer molecules) and external lubricants (to prevent sticking between the hot compound and the metal machinery), although this distinction may simply indicate the degree of compatibility of the lubricant with the melt.

Pigments, both inorganic and organic, are widely used for colouration, and may also be used to improve light stability and weather-fastness.

Fillers are used primarily to reduce cost, and the main ones are calcium carbonate and kaolin. They can also be used to enhance specific properties, e.g. electrical conduction, flame resistance, surface finish, etc.

Impact modifiers are rubbery polymers that are dispersed in the compound, and they function by blocking the path of crack propagation; there are numerous types of such compound.

In addition to the above, there are various other additives that may be included, such as processing aids, fungicides, flame retardants, antistatic agents, UV absorbers, or blowing agents.

2.4.2 Mixing

The PVC powder is mixed with the other constituents to produce a homogeneous blend before the actual processing. This is a vitally important stage, since inefficient mixing and compounding will result in a defective or otherwise poor quality product. The compounding can be used to produce solid pellets of material that can be stored and processed at a later stage into the final product; in fact, blended PVC compounds are commercially available in addition to the pure polymer. It is also quite common, however, for companies to do their own in-house compounding where the raw materials are mixed, compounded, and processed in continuous stages.

The mixing can be done in various ways, depending on the formulation and processing method to be used. In solid phase mixing, the constituents (added in a particular order for best results) are fed into a high speed turbine impeller mixer. The blades create a homogeneous mixture and the friction generates considerable heat; this results in a "dryblend" powder that is available for further processing, even if some of the constituents (such as stabiliser) are originally in liquid form. External heat can also be applied in some blending processes, so that the mix gels.

Subsequent to mixing, an extra homogenisation and compounding is obtained by subjecting the mix to a combination of heat and shear, for example using a screw-extruder. The hot gel can then either be pelletised or pass to the next stage of processing.

Fluid phase mixing is used to produce PVC pastes, which is a generic term used for mixtures that are viscous liquids or gels at room temperature, and can thus be moulded or spread as desired before being hardened at a higher temperature. The pastes are prepared by mixing the PVC powder (usually of high molecular weight) and other solid additives in the liquid components of the formulation, such as the plasticiser. This is generally done in a vat with a low-speed impeller. Pastes where the only liquid component is the plasticiser are termed plastisols, while organosols contain other organic solvents as thinners. Plastigels are plastisols to which thickeners such as aluminium stearate have been added to form a putty-like consistency; aqueous PVC dispersions, termed hydrosols, can also be produced for coating certain absorbent substrates.

2.4.3   Processing

(i)   Calendering. PVC sheet is produced principally by the calendering process, although some rigid profiles are made by extrusion. The sheet thus produced, encompassing a wide range of colours, gauges, and mechanical characteristics, acts as the starting feedstock for a vast number of thermoformed products used in many industries. Organotin stabilisers (principally thiotins) are very widely used in calendered sheet, usually incorporated at a level of 1.0-1.5 parts per hundred of resin (phr). West Germany, for example, is the major European producer with approximately 100,000 tonnes annually in the early 1980s, virtually all of it tin-stabilised. The clarity imparted to the clear product by organotins naturally means that their main application is for this particular material; however, they are also used in opaque and pigmented sheet by many manufacturers, which may be typically converted into stationery bindings, tapes, printed displays, etc. Butyltin stabilisers may be used for technical grades of PVC, while octyltins are used for those grades required for food-contact applications. However for the manufacturer who is building up a stock of various types of calendered sheet, it is needlessly time-consuming to change the PVC formulation more often than is necessary, and it is more economical and efficient to produce a PVC grade that is suitable for any end-use that may be required. Thus it is fairly common for calendrers to formulate all their products as being food-grade material, whether or not it is going to be used for this application.

Calendering is one of the earliest processing methods, used originally in the rubber industry, but has been continually refined until today it is able to meet the demands of mass production of high quality sheet to very close tolerances, both for plasticised and rigid product.

Fig. 2.3a. A section of PVC is taken from the rolling mill and fed to the calender.
Courtesy: Stanley Smith and Co. Plastics Ltd., Isleworth, U.K.

Fig. 2.3b. The clear sheet emerges from the calender and passes around the cooling rollers, shown at left.
Courtesy: Stanley Smith and Co. Plastics Ltd., Isleworth, U.K.

The various stages of the manufacturing process, as with the other processing methods, may vary in detail or layout between different manufacturers, but are broadly similar. The first stage is the mixing of the various components at about 70°-80°C by a high speed mixer; generally a low-to-medium molecular weight PVC is used for rigid sheet, and mass-polymer is often used for the clear colourless type due to there being less colour development during processing. The mix is put through an extruder which converts it to a hot homogeneous gel; this is carried to a rolling mill which compresses it into a viscous mass. A narrow section of material is continuously taken from this and fed to the nip between the first two calender bowls by a feeding arm moving to and fro. Most calenders nowadays consist of four cylindrical rollers or bowls, which can be arranged in a number of ways; the most common are the "inverted L" and "Z" configurations.

 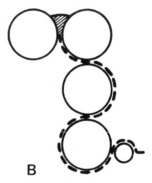

A

B

Fig. 2.4. Typical configurations of calender bowls : (A) Z type, (B) Inverted L type.

The bowls can be up to 2 m wide and must have highly polished and flawless metal surfaces, since any defects will be transferred to the PVC sheet. The bowls are internally heated (usually by oil) to about 200°C for rigid PVC, but each bowl is maintained at a different temperature, and they are individually driven at different speeds. This individual control allows materials with different rheological properties to be processed on the same machine. Other geometrical and pressure adjustments can also be made, enabling uniform sheet to be produced to close tolerances from materials with differing mechanical characteristics.

The bowls squeeze out the hot material into a thin continuous sheet, which must be cooled before it is taken up on a final spool; this is done by passing it around a series of water-cooled rollers. It is important that the material is not stretched at this stage, otherwise the final sheet will regain its former dimensions if it becomes heated up again, i.e. it will shrink. It should be noted however that heat-shrinkable film is an important sector of the packaging market, and this type of material can be produced by deliberately stretching the film at fairly low temperatures before it cools.

Calendered sheet is produced in a range of gauges from approximately 75 to 600 μm. Embossing and/or roller gravure printing is often also carried out in the calender train, or as separate operations subsequently.

(ii) Extrusion. Extrusion is used to produce a great variety of articles in PVC, and the extrusion head can be designed for pipes, tubes, filaments, sheeting, complex profiles, wall cladding, etc. A blow-head fitted to the extruder can be employed to produce hollow articles and blown film.

The earliest type of extruder was the single-screw machine, and these are still widely used although there are an increasing number of twin-screw and multi-screw systems as well. The single-screw machine basically consists of a horizontal steel tube or barrel with a hard mirror-polished interior lining. This barrel houses a rotatable screw, usually of diameter between 30 mm and 250 mm. A dryblended PVC compound, either as powder or pellets, is fed in at one end from a hopper, and this is picked up by the rotating screw; generally, a medium-to-high molecular weight PVC is used for extrusion. The PVC is successively compressed, softened and homogenised as it passes down the barrel, being subjected to external heat and great shear. Rigid PVC compounds naturally require higher temperatures for this process, of around 200°C, and efficient heat stabilisation and lubrication is essential. The melt is forced through an extrusion head under high pressure, and the extruded profile is air- or water-cooled after it emerges.

Twin-screw extruders comprise two parallel screws side by side in a horizontal plane, which can rotate either in the same or opposite directions. Material is conveyed by passing from one screw to another, and the through-put is controlled by the rate of resin feed, whereas for single-screw machines, it is controlled by the screw speed. The more recently introduced multi-screw extruders may have as many as four screws arranged in various ways, and are used for speciality applications such as large diameter pipes. Twin and multi-screw machines can be used for high through-puts, and they can operate at lower temperatures than single-screw systems, thus lower levels or less efficient heat stabilisers may be acceptable here.

Organotins find a substantial outlet in extruded parts, although the PVC formulations used (including the stabilisers) can vary considerably between different countries. In the United States, for example, there is a long tradition of using tin stabilisers for many types of extrusion (e.g. piping), while lead compounds predominate in European pipe manufacture. Butyl thiotins are commonly used in the U.S.A., and methyltins have also been widely accepted by U.S. manufacturers because of their exceptional heat stabilising action and consequent cost-effectiveness. Estertin stabilisers are employed as well, and some formulations are claimed to be specifically designed for extrusions. Single-, twin- and multi-screw extruders are all being employed for PVC piping, and the level of stabiliser required depends on the process used. In single-screw machines, 1-2 phr of organotin stabiliser will be effective for the temperatures generated, while only 0.5 phr may be required in multi-screw equipment. Some methyltins are claimed to be effective at levels as low as 0.85 phr in single-screw, and 0.25 phr in twin-screw extruders. These low levels make the compounds very cost-competitive with alternatives, and as mentioned earlier, some mixed-metal systems such as calcium-, barium-, and antimony-tin have also been introduced in the U.S.A. which are claimed to cut costs further.

Profiled transparent rooflights and cladding are other extrusions that employ organotins, generally maleates used at a level of 2.5-3.0 phr. These applications are discussed in Section 2.5.

(iii) Injection moulding. In this process, the PVC compound is first heated and sheared into a homogeneous melt, which is then injected under high pressure into a mould, resulting in a finished part that is removed from the mould when it has cooled. There are two types of injection moulding system in use, namely the screw-preplasticator and reciprocating screw systems. In the first, a fixed-screw extruder plasticises the PVC compound and continuously feeds it into an injection plunger. When the desired amount of material has accumulated, it is rapidly injected into the mould by a separate plunger system. In the second, more recent, system, a screw extruder plasticises the compound as before, and the melt is forced in front of the screw via a one-way valve. As the melt accumulates, the screw is forced back along the barrel until the desired quantity of compound is ready for injection. Then the screw is powered forward, injecting the melt into the mould. This type of system has made the injection moulding of rigid PVC possible, and opened up new areas of application.

Organotins are not used to any great extent in injection moulding, but are employed for some articles such as pipe-fittings.

38

(iv) <u>Blow moulding</u>. Blow moulding is a fast efficient process for the mass-
production of hollow containers, and has been a steadily growing market for PVC
as well as other plastics. The process basically consists of forming a PVC preform or parison for the required container; this is positioned inside a mould
and injected with air, so that the parison is blown up like a balloon to fill
the mould. There are two main techniques for blow moulding, depending on how
the parison is produced, plus the newer development of stretch-blow moulding
which is used to make PVC bottles having particular permeability properties.

In extrusion-blow moulding, the PVC compound (as either powder or pellets) is
fed into a single-screw extruder, and the parison material is continuously extruded as a hot homogeneous tube. A two-part mould closes around a length of
this tube, cutting it off and sealing one end by crimping it together. Air is
then injected into the open end, blowing up the parison to the mould shape.
After cooling, the extraneous "flash" material is removed from the container,
and it is ejected. A single extruder head can feed a number of moulds that are
sequentially positioned in front of it, and there are various more complex
arrangements possible, such as multiple die heads, to increase output. Extrusion-blow moulding can produce a large range of container sizes, although the
bottom seam is a possible source of weakness if it is subjected to stress. The
technique has become increasingly sophisticated with regard to control of

Fig. 2.5. A bottle produced by extrusion blow-moulding.
Courtesy: Leon Frenkel Ltd., Kent, U.K.

operating parameters, and the parison is programmed to a particular shape as it is extruded in order to give uniform material distribution in the final product.

In injection-blow moulding, the PVC parison is produced by injection moulding it in a cavity, the parison (including the integral screw neck) being formed over a steel rod or mandrel. This rod with the parison is transferred to the container mould, where the full bottle is blown. Since the parison shape is determined exactly each time by the injection mould, containers can be precisely reproduced quite easily with no waste, and these containers have no seams as possible weak points. However, this process is limited to containers up to 0.5 litre capacity. The manufacturing process is commonly carried out on a three-station rotatory system, where the respective positions correspond to injection mould, blow mould, and bottle eject.

Although PVC bottles manufactured by the above process possess adequate mechanical strength, they are not impermeable enough to gases (specifically, carbon dioxide) to be suitable for carbonated drinks. However, improved gas impermeability, together with other properties, can now be achieved by the process of

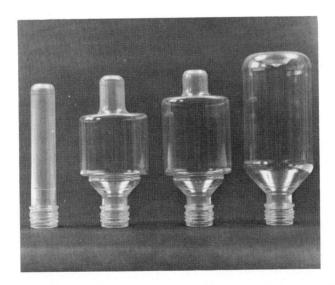

Fig. 2.6. Stages of production of an injection blow-stretch moulded PVC bottle.

biaxial orientation of PVC, i.e. the controlled "stretching" of the PVC in two perpendicular directions. This technique is now applied to a range of polymers using a number of different manufacturing processes. For PVC, an injection-blow-stretch system (IBS) has been devised, the first part of which is similar to conventional injection blow moulding. The PVC compound is fed to a screw extruder, which feeds it to the injection mould. The preform is produced on a steel rod at the first station, and this then passes sequentially through a number of conditioning stations that reduce the preform temperature to 90°C before it reaches the blow-moulding stage. At this temperature, the PVC is in a thermo-elastic state, and the polymer chains are less mobile; more force is required to stretch the polymer than in conventional blow moulding. Air is injected under high pressure, and the polymer expansion is also aided mechanically. The relationship between the dimensions of the preform and those of the final blown bottle is carefully controlled to produce the required degree of biaxial orientation. The bottle is allowed to cool in the mould to preserve dimensional stability and to freeze the stresses that have been produced, before being ejected. The resulting bottle has a considerably reduced gas permeability, as well as increased impact strength and compression resistance. A very high clarity and freedom from striations is also said to result, and such bottles have a high strength-to-weight ratio.

Blow-moulded bottles represent an important sector for organotin stabilisers, where they are used at a level of 1.0-1.5 phr. Octyl thiotins are the most common type, but methyl and ester thiotins are also used.

(v)   Other techniques. The methods described so far are the principal ones used for processing PVC resin, but there are numerous others, some of which will be considered in the next section, where they are relevant to tin-stabilised PVC. There are various methods for processing PVC pastes, for example; these include roll-coating, rotational casting, dipping, and slush moulding.

2.5   APPLICATIONS OF TIN-STABILISED PVC
2.5.1   Benefits of using organotins
The particular advantages conferred by organotin stabilisers to both the manufacturing process and the product can be summarised as follows:

(i)   The most important benefit is their very efficient heat stabilising action, and the thiotins in particular are the most effective known heat stabilisers. This is especially required for rigid PVC, which needs higher processing temperatures than the plasticised form.

(ii)   A high degree of clarity is conferred to transparent PVC, which is a great advantage in many packaging and building applications.

(iii)   Unlike other types of stabiliser, the organotins have a wide spectrum of use, and are effective in all grades of PVC, including suspension-, emulsion-, and mass-polymerised types. They are also effective in many copolymers, and also in polyblends or polymer mixtures. This will obviously be advantageous to the processor who handles several different types of PVC polymer from several sources.

(iv)   They have good compatibility with the other common components of the PVC compound, e.g. lubricants, plasticiser, impact modifier, pigment, etc. This facilitates processing and minimises possible manufacturing problems such as the "plate out" of incompatible additives on the machinery, thus reducing the down-time of the equipment and improving the economics of the process.

(v)   The low toxicity and good leach-resistance of certain types of stabiliser enables them to be used in food-contact rigid PVC and other sensitive applications. (In plasticised PVC, the leach-resistance is insufficient for use in this field). As stated earlier, di-n-octyltin diisooctylthioglycolate and maleate have approval in many countries, including the United States and most of Europe. In the Federal Republic of Germany, the B.G.A. has approved the use of di-n-octyltin compounds of lauric acid, maleic acid, maleic acid esters, and thioglycolic acid esters. Governmental approval is usually subject to certain conditions, and in F. R. Germany, for example, the concentration of these stabilisers must not exceed 1.5%, and the PVC must not contain more than a total of 1.5% of certain other additives such as silicone fluids, liquid paraffins, esters, or certain oils. The B.G.A. has also approved the use of monobutyltin sulphide, dimethyltin diisooctylthioglycolate, and di-estertin diisooctylthioglycolate. (These approvals also include the monoorganotin compounds that are often used synergistically with the diorganotin analogues). The last two types of stabiliser are also approved in the U.S.A. for use in potable water piping.

(vi)   The sulphur-free organotins confer good light-stability, making possible a range of outdoor applications.

(vii)   The use of organotins eliminates the risk of discolouration by hydrogen sulphide in the atmosphere, or by sulphur-containing chemicals; other heavy-metal stabilisers may become discoloured in these conditions.

The drawbacks of organotin stabilisers can be summarised as:

a)   They are relatively costly compared to other stabilisers, although this can be offset to some extent by their generally lower loading levels.

b)   The mercaptides or thiotins have a characteristic odour which is often noticeable during processing, although rarely in the final product. Some maleate-
type stabilisers may give rise to lachrymatory by-products during processing,
but this has no effect on the finished article.

c)   They do not confer any lubricity to the PVC compound, and thus efficient
lubrication may be required.

d)   The mercaptides are difficult to process with flexible compounds due to the
development of excessive adhesion to equipment surfaces, and they are primarily
used for unplasticised PVC.

We shall now look at the various areas of application for tin-stabilised PVC.

2.5.2   Packaging

Packaging represents a major end-use for organotin stabilisers, in the forms
of both calendered sheet and blow-moulded bottles. There is an almost unlimited
variety of goods that are marketed in PVC or part-PVC containers, from foodstuffs and beverages to household goods, giftware, cosmetics, pharmaceuticals,
and so forth.

Calendered PVC sheet is commonly converted into the desired shape by vacuum-
forming, and a wide range of commercial machinery is available for this purpose,
from small hand-operated units to very large, fully automated models for mass
production. Their mode of operation is essentially similar: a section of PVC
sheet is drawn from a supply roll and clamped horizontally by its edges in a
frame. Below the sheet is the desired mould, usually made of aluminium; this
contains a series of narrow holes that allow air to be pumped out through the
mould. A hood containing radiant heating elements is drawn across the PVC to
heat it up to a thermoplastic temperature, and after a suitable time, this is
withdrawn and the mould on its table is raised to contact the underside of the
sheet. On activating a vacuum pump, the air is pumped out of the chamber and
the sheet is drawn tightly around the mould, reproducing all the contours precisely. The vacuum is released, and air blowers cool the sheet, which is then
removed.

This technique is used to produce "blister-packs", which are a common way of
packaging fairly small items for sale; a blister pack comprises a moulded shell
of rigid clear PVC on a card backing, and is a convenient way of displaying and
advertising an article to the best advantage. A set of perhaps twelve blisters
will be moulded on a single sheet, which is then bonded to the card backing with

Fig. 2.7. Clear PVC forms an ideal display package for many types of product, including foods, cosmetics, and giftware.
Photograph courtesy: Klöckner Pentaplast Ltd., Reading, U.K.

the appropriate articles in position. The sheet is then cut up into individual packs.

Another type of display package is the "skin pack", where the article is held to a card backing by a tight, barely-visible film of material. Here, the article itself on the backing card acts as the mould in the vacuum-forming process, and the card contains perforations so that air can be pumped out to draw the clear film tightly around the article.

Another thermoforming technique employed is "positive pressure forming". Instead of a vacuum, this method uses a jet of air pressure to force hot PVC sheet into a hollow mould. This is used for the mass-production of carbons, tubs, etc.

Calendered sheet is being increasingly adapted to the needs of modern packaging, and grades of PVC are now offered that have good anti-static and de-nesting properties (i.e. the removal of containers when they are stacked one inside the other), as well as a degree of resistance to heat, deep-freeze cold, and gamma-sterilisation radiation. With regard to the last property, it is interesting to note that an organotin stabiliser has recently been introduced

(M and T's Gammashield *) which contains an additive designed to protect PVC against sterilisation radiation.

Good barrier properties to water vapour and oxygen are an essential attribute for packaging materials designed for long-term storage of sensitive products such as foods or pharmaceuticals. While grades of PVC are marketed with superior barrier properties, it is becoming increasingly common to employ multi-layer laminates to greatly reduce permeability. An important barrier film is polyvinylidene chloride (PVdC), which can be applied to a substrate in various ways,

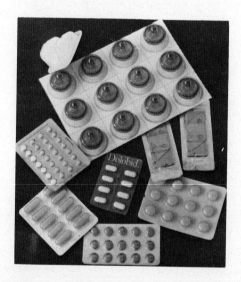

Fig. 2.8. Pharmaceutical blister-packs in PVC.
Photograph courtesy: Klöckner Pentaplast Ltd., Reading, U.K.

such as pressure-laminating a commercially available PVdC film (e.g. Dow's Saran *). These high-barrier multi-layer films, which can incorporate a wide range of materials, including PVC, will be a rapid growth area in packaging in the future, and will tend to reduce the need for energy-consuming refrigeration for various food products.

Blow-moulded clear <u>PVC bottles</u> have now become well-established for various liquid products such as fruit squashes, cooking oils, mineral waters, vinegar, as well as various toiletries. They have taken a large part of the traditional markets from glass, where PVC and other plastics have the obvious advantages

* Trade name throughout.

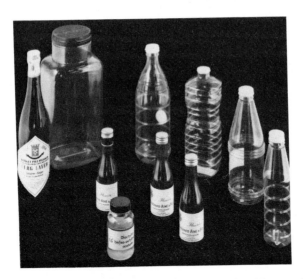

Fig. 2.9. Dioctyltin stabilisers are commonly used for blow-moulded bottles and jars for food use.

over glass of being lightweight and shatterproof. These benefits become more marked with larger containers, and 1.5 l and 2 l bottles are manageable for the consumer when produced in PVC or other plastics, whereas they would be heavy and unwieldy if made of glass. Large sweet jars, for example, are now mainly in PVC in the U.K., having formerly been entirely in glass. It is interesting to note that the glass bottle manufacturers have introduced heat-shrink plastics sleeves for bottles which combine a protective shield for the glass with all-round decoration and labelling; one such commercial protective sleeve is in a tin-stabilised PVC.

As described earlier, the technique of producing bottles in biaxially-orientated PVC has enabled carbonated beverages to be added to the range of drinks. Some bottle designs are reported by the manufacturers to be suitable for drinks containing up to 8 g/l carbon dioxide. It should be noted that drinks such as beer and cider require a high oxygen barrier to prevent their deterioration, and the biaxial-orientation process alone does not provide sufficient impermeability to oxygen. For this area, a PVdC barrier layer is required on the container, and these have been introduced for PET recently, although not as yet for PVC.

Blow-moulded bottles represent an important sector for organotin stabilisers, as mentioned earlier, but the pattern of usage varies markedly between countries. In F. R. Germany and the U.K., most bottles are tin-stabilised, but in the very large French blow-moulding industry, calcium-zinc stabilisers predominate.

2.5.3   Building

The construction industry represents a large outlet for PVC, particularly in the U.S.A. where it consumed 1.09 m tonnes in 1982, representing about 45% of the total U.S. consumption of PVC. In the U.S.A., PVC is widely used for <u>cladding</u> and <u>siding</u> in buildings, as well as for many applications of <u>piping</u> related to this industry; these include drain, waste and vent systems, water and sewage mains, gas distribution, electrical and communications conduits, and industrial process piping.

Fig. 2.10. Rigid PVC cladding for buildings is a widespread use in the U.S.A. Photograph courtesy: M and T Chemicals Inc., Rahway, U.S.A.

PVC piping is light, easily joined, resistant to chemical attack and corrosion, of high impact strength, and does not require painting; its use worldwide seems likely to increase in the forseeable future.

While the use of extruded wall-cladding as an external facing material is well-established in the U.S.A., it is less common in Europe. However, PVC cladding in Europe is commonly used as an internal facing material in many commercial and civic structures. Imperial Chemical Industries have been producing their Darvic * range of rigid PVC sheet, for example, for over 20 years. This product is made not by extrusion but by calendering the PVC into thin sheets,

Fig. 2.11. PVC piping is easy to install and is suitable for many industrial
applications.
Photograph courtesy: Celanese Plastics Co., Ohio, U.S.A.

which are subsequently laminated together using heat and pressure. Organotin
stabilisers are employed in the wide range of sheet products made by ICI;
Darvic, for example, is produced in a standard grade in fifteen colours (approved for food-contact use), an industrial grade, a transparent grade, a clear
wire-laminate grade, and an opal grade. These materials are claimed to possess
excellent mechanical properties with high resistance to weathering and chemical
attack. Another important factor, as with all PVC that is used for construction
purposes, is its good fire-test rating, where it is rated in the lowest category
of flammability according to BS 2782 Part 508A. This type of cladding is employed for walls and ceilings in food shops, stores, and markets, and for fit-
tings such as counter-tops, fascia, and shelving. Associated uses include linings for transit containers (e.g. refrigerated transporters) and trays for un-
wrapped foodstuffs. Large civic structures also employ internal PVC cladding,
one recent example being the new Clyde Tunnel at Glasgow.

Transparent PVC roofing material is one building application that is quite
widespread in the U.K., although it seems to be fairly specific to this country.
This product usually employs dibutyltin maleate stabiliser due to its endowing
better light stability and weathering properties to the material, and additional
UV stabilisers are generally also used. ICI's Sintilon * is a profiled rooflight sheet produced by extrusion, and is claimed to maintain good weather protection for at least 20 years. It has also been shown [36] that the use of PVC

Fig. 2.12. PVC internal cladding in the Clyde Tunnel.
Photograph courtesy: Imperial Chemical Industries plc, London, U.K.

Fig. 2.13. The use of organotin stabilisers in profiled rooflight sheet endows
high clarity and good weatherability.
Photograph courtesy: ICI plc, London, U.K.

rooflights may actually assist in fire-fighting by acting as automatic vents when a fire approaches; due to its low softening point (72-74°C), the rooflight opens up as the heat builds up, releasing the heat and smoke outwards, while no burning droplets are produced by the material.

As a development from Sintilon, a profiled PVC material Sinticlad * has recently been introduced. This extruded cladding material has been developed to meet the needs of particularly corrosive environments, and is used in the construction of, for example, chemical plants.

Fig. 2.14. Profiled cladding can also be designed for use in corrosive atmospheres, as in this chemical plant in Ayrshire.
Photograph courtesy: ICI plc, London, U.K.

One other area in the building industry where PVC extrusions have been growing steadily is window profiles. This is an important sector in parts of Europe, particularly F. R. Germany where in 1979 an estimated 35% of the window frames made were in PVC; as yet, this market is small in the U.K. Tin stabilisers are not yet used to any great extent in Europe for this product, where barium/cadmium or lead stabilisers are generally used instead. However, several companies are known to have developed organotin-based formulations that have overcome earlier problems of processability and long-term heat stability associated with tin maleates in this application. Ciba-Geigy, for example, has recently introduced Irgastab * T633, a liquid butyltin carboxylate/mercaptide stabiliser designed for outdoor rigid pigmented PVC such as window profiles. This is claimed to give good processability, freedom from plate-out, and good long-term stability and weatherability; it is used at concentrations of 2.0-2.5 phr in twin-screw extruders for profiles, or 2.5-3.0 phr for sheet. The use of organotins in this field may increase if there are future legislative restrictions on cadmium compounds.

2.5.4   Other areas

While packaging and construction represent the major sectors for organotin stabilisers, there are a number of other areas where they are employed.

The use of <u>vinyl wallcoverings</u> has been a rapidly expanding one since they were introduced around 1965, particularly in the U.K. where a lot of the development work was done. Some wallcoverings are produced by laminating calendered PVC sheet to a backing paper, but a more efficient and versatile method uses PVC paste as the basis material. The paste is made up by mixing together the PVC polymer powder, plasticiser, stabiliser, filler/pigment (titanium dioxide), extender, and sometimes adding a blowing agent (commonly azodicarbonamide) and colour pigment. These are homogenised using a rotary impeller, and the fluid mixture is then fed to a coating machine that spreads it evenly on a backing paper; the technique used is "reverse roll coating", where the actual coating roller moves in the opposite direction to the flow of the paper. The coating weight is controlled by the gap between the metering roll and the coating roll.

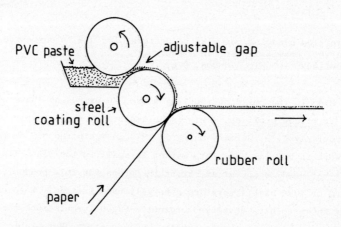

Fig. 2.15a. The reverse roll coating technique for wallcoverings.

Fig. 2.15b. Stages in the production of tin-stabilised vinyl wallcoverings.
Courtesy: Nairn Coated Products Ltd., Lancaster, U.K.

The coated paper or web then passes through a convection oven that raises its
temperature to about 200°C, which cures the plastisol to a solid coating. A
major British manufacturer, having investigated different stabiliser systems,
chose one based on dioctyltin maleate as the most suitable, since it is efficient, non-toxic, non-odorous, and works equally well with PVC from several commercial sources; the level of incorporation required is very low (less than 1
phr), making it as cost-effective as any other system.

The basic white web material produced is subsequently decorated and finished
to create a vast range of patterned wallcoverings. Printing is done by a rol-
ler-gravure method, and up to six colour designs can be superimposed, each ink
being cured by hot air before the next is applied. This may be followed by embossing, where the pre-heated printed web is passed over a water-cooled engraved
metal roller which imprints its pattern into the PVC coating. Another interesting technique which can be used instead is "chemical embossing", which is applied to web material containing blowing agent (e.g. azodicarbonamide, which decomposes at about 200°C); this has been initially cured at only 140°C so as not
to decompose the latter. However, on passing the printed web through a second
oven at 200°C, the PVC is made to foam and becomes slightly raised, although the
surface remains intact. One chemical embossing process depends on the fact that
certain of the printing inks applied contain inhibitors that prevent the blowing
agent decomposing at this temperature. Thus the PVC coating is in raised

relief everywhere except where these inks are located, creating an embossed effect that is furthermore always in exact registration with the printed design.

A final effect that may be applied is flocking. The flock pattern is applied to the printed web by first screen-printing adhesive in the required design from a perforated nickel cylinder. The flocking material (rayon fibres with a special conductive coating) is applied to the web electrostatically by charging it up to about 70,000 volts, the web being earthed; the fibres stick vertically in the adhesive, but are repelled away from non-adhesive areas. The adhesive is finally cured in a hot oven to leave the flock design firmly in place.

The production of vinyl wallcoverings thus involves several heating cycles, and efficient stabilisation is essential particularly in view of the exact colour-matching required in the product.

A similar kind of PVC paste technology is used for vinyl floor-coverings, although of course this material has to be much tougher, and the backing material is a fabric or polymer. The actual patterned vinyl layer has a tough topmost coating of clear vinyl in order to protect it from wear. Organotin stabilisers are not generally used for the patterned layer, but are used in the paste formulation for the top clear layer by British manufacturers.

Gramophone records are a well-established and fairly substantial market for PVC amounting to 115,000 tons of resin in the U.S.A. and W. Europe in 1982; the actual polymer used is a copolymer of vinyl chloride and vinyl acetate. The formulations employed again vary between countries and companies, but butyltin stabilisers are used in F. R. Germany for this product. Records are manufactured by a compression moulding technique where the hot plastic homogenised compound is pressed between two metal stampers. A number of companies are also known to be studying the possibility of using organotins in video discs, where requirements for the discs place rigorous demands on processing.

Compression moulding applied to PVC sheet is used to produce various goods, including credit cards. This market consumed 9,000 tonnes of PVC in the U.S.A. in 1982, all of it tin-stabilised.

PVC is often laminated with other materials to form useful products, some examples of which have already been described. In the U.K., one particular application is coil-coated steel, which also uses tin-stabilised PVC.

PVC cellular materials or foams are well-established, particularly the plasticised types employed for coated fabrics and wall- and floor-coverings, the latter having just been discussed. More recently, there has been rapid development in the production of extruded rigid cellular profiles. These include piping and thick complex-section profiles that are used as a replacement for wood in the construction industry. These rigid extruded foams also employ a chemical blowing agent such as azodicarbonamide, and it is interesting that the heat-stabilisers used in the PVC compound have a dual role; in addition to providing heat stability, they also act as "kickers" for the blowing agent, i.e. they lower its decomposition temperature. This catalytic effect occurs with several metal salts, and modified organotin stabilisers have been developed which can act in this way [37]. They are used for this type of product, particularly in the U.S.A., and this sector is expected to grow steadily in the future.

2.6   SUPPLIERS OF ORGANOTIN STABILISERS

Listed in this section are the major manufacturers and suppliers of organotin stabilisers, together with the trade names used by the companies for their product. Of course, the products available change periodically as some are withdrawn and new ones appear, and thus a comprehensive list of individual formulations would be soon out-dated.

The general types of product available from each company are briefly summarised in the table, and detailed information can be obtained from the Head Office address shown; many of the companies also have other offices and agents world-wide.

TABLE 2.3

Some major suppliers of organotin stabilisers

Company	Product Trade Name	Types of Product
Acima Chemical Industries Ltd., Inc., P.O. Box, CH-9470 Buchs 1 SG, Switzerland.	Metastab	Dibutyltin dilaurate, mercaptides and carboxylates; dioctyltin dilaurate, mercaptides, and maleates.
Akzo Chemie bv, Stationsstraat 48, 3800 AZ Amersfoort, The Netherlands.	Stanclere	Estertin mercaptides, octyltin mercaptides, butyltin carboxylates and mercaptides.

Company	Product Trade Name	Types of Product
Argus Chemical Corporation, 633, Court Street, Brooklyn, NY 11231, U.S.A.	Mark	Octyl- and butyltin mercaptides and maleates.
Otto Bärlocher GmbH, Riesstrasse 16, Postfach 500108, D-8000 München 50, F. R. Germany.	Okstan	Dibutyltin mercaptides and dilaurate, dialkyltin maleic acid semi-ester, dioctyltin mercaptides.
Carstab Corporation, West Street, Cincinnati, OH 45215, U.S.A.	Advastab	Methyltin mercaptides.
Ciba-Geigy Marienberg GmbH, Postfach 209, D-6140 Bensheim, F. R. Germany.	Irgastab	Dibutyltin maleates and mercaptides, butyltin carboxylate/mercaptides, dioctyltin thioglycolates, estertin mercaptides.
Diamond Shamrock Europe Ltd., P.O. Box 1, Silk Street, Eccles, Manchester M30 0BH, U.K.	Lankromark	Butyltin carboxylates and mercaptides, octyltin mercaptides.
Hoechst A.G., Postfach 101567, D-8900 Augsburg 1, F. R. Germany.	Hostastab	Dibutyltin thioglycolates, carboxylates, thioglycolate/carboxylate; dioctyltin thioglycolates.
M and T Chemicals Inc., Woodbridge Road, P.O. Box 1104, Rahway, NJ 07065, U.S.A.	Thermolite	Dibutyltin carboxylates and mercaptides; dioctyltin maleate and isooctylthioglycolate.
Nitto Kasei Co. Ltd., 369, Nishiawaji-cho, 2-Chome, Osaka 533, Japan.	TVS	Butyl- and octyltin mercaptides, maleates, and laurates.
Polytitan S.A., Boite Postale 2, 60301 Senlis Cedex, France.	Stannofix	Octyl- and butyltin mercaptides and carboxylates.
Swedstab AB, Box 69, S-265 01 Åstorp, Sweden.	Swedstab	Dibutyltin diisooctylthioglycolate, ditallate, dilaurate, maleate; dioctyltin mercaptides.

Company	Product Trade Name	Types of Product
VEB Chemiekombinat Bitterfeld, Bitterfeld 44, German Democratic Republic.	BTS, OTS, BT, BTL	Dibutyltin and dioctyltin mercaptides, dibutyltin maleate and dilaurate.

REFERENCES

1   C.W. Johnston, in L.I. Nass (Ed.), Encyclopaedia of PVC, Marcel Dekker, New York, 1976.
2   R.H. Burgess, in A. Whelan and J.L. Craft (Eds.), Developments in PVC Production and Processing - I, Applied Science Publishers, London, 1977.
3   H.M. Clayton, ibid.
4   H.M. Clayton, ibid.
5   E.W. Rugeley and W. Quattlebaum, U.S. Pat. 2,344,002 (1944).
6   W. Quattlebaum and C. Noffsinger, U.S. Pat. 2,307,157; GB Pat. 557,477 (1943).
7   E.g., D. Cleverdon and J. Staudinger, GB Pat. 590,734; U.S. Pat. 2,481,086 (1947).
8   Burt, U.S. Pat. 2,489,518; GB Pat. 679,655 (1949).
9   E.g., J. Church, E. Johnson and H. Ramsden, U.S. Pat. 2,591,675; GB Pat. 718,393 (1952).
10 E.g., J. Church, H. Ramsden, H. Hirschland and H. Buchanan, U.S. Pat. 2,743,257 (1956).
11 R. Schlattman, U.S. Pat. 2,455,613 (1948).
12 G. Mack and E. Parker, U.S. Pat. 2,634,281 (1953).
13 W. Leistner and A. Hecker, U.S. Pat. 2,680,107 (1954).
14 C.E. Best, U.S. Pat. 2,731,484; GB Pat. 728,953 (1956).
15 H. Verity Smith, International Tin Research Institute Publication No. 302, 1959.
16 J.G.A. Luijten and S. Pezarro, British Plastics, 30 (1957) 183.
17 U. Thust, Tin and its Uses, No. 122 (1979) 3.
18 H. Verity Smith, British Plastics, Aug. 1964, 445.
19 P. Klimsch and P. Kühnert, Plaste u. Kauc., 16 (1969) 242.
20 Federal Health Dept. Report 33, Bundesgesundheitsblatt 9 (1966) 322.
21 R. Schlimper, Plaste u. Kautschuk, 14 (1967) 657.
22 Z. Mayer, J. Macromol. Sci., (C), 10 (1974) 263.
23 D. Braun and W. Quarg, Angew. Makromol. Chem., 29/30 (1973) 163.
24 D. Braun and M. Thallmaier, Kunststoffe, 56 (1966) 80.
25 D. Druesedow and C.F. Gibbs, Mod. Plastics, 30 (10) (1953) 123.
26 K.S. Minsker, M.I. Abdullin, S.V. Kolesov and G. E. Zaikov, in G. Scott (Ed.), Developments in Polymer Stabilisation - 6, Applied Science Publishers, London, 1983, pp. 173-238.
27 D. Braun and R.F. Bender, Eur. Polym. J., Supp., (1969) 269.
28 R.M. Aseeva, J.G. Aseev, A.A. Berlin and A.A. Kasatockin, Zhurn. Strukturnoy Chim., 6 (1) (1965) 47.
29 W.H. Gibb and J.R. MacCallum, Eur. Polym. J., 7 (1971) 1231.
30 Idem, ibid., 8 (1972) 1223.
31 Idem, ibid., 10 (1974) 529.
32 H.O. Wirth and H. Andreas, Pure and App. Chem., 49 (1977) 627.
33 A. Frye, R.W. Horst and M.A. Paliobagis, J. Polym. Sci., A2 (1964) 1765, 1785, 1801.
34 Manual of PVC Additives, Ciba-Geigy Marienberg GmbH, 1971, p.33.
35 W. Jasching, Kunststoffe, 52 (1962) 458.
36 PVC Rooflights for Venting Fires in Single Storey Buildings, Fire Research Station Technical Paper No. 14.
37 R.D. Dworkin, Trans. 35th Soc. Plast. Engineers Annual Tech. Conf., Montreal, April, 1977.

Chapter 3

CATALYSIS

While the major usage of dialkyltin compounds is in the stabilisation of PVC, they, along with certain other organic and inorganic compounds of tin, also have important uses as catalysts in three main areas. These are i) polyurethane production, ii) silicone curing agents, and iii) esterification reactions. These will be looked at in turn.

3.1 POLYURETHANES

3.1.1 Introduction

The great expansion in the production and applications of man-made polymers has been a prominent feature of our lives since 1945, and in almost every conceivable area, uses have been found for new materials that are able to equal or surpass the performance of traditional ones. One such group of materials that has undergone rapid growth are the polyurethanes; the term "group of materials" is rather more appropriate here since the physical appearance and characteristics of polyurethanes can range from lightweight flexible foams to rigid high-strength engineering polymers, via tough elastomeric compounds. Unlike PVC, this versatility is not solely the result of incorporating numerous additives with the basic polymer, but is controlled in the main by the actual compounds that are polymerised together. Thus there is no such thing as a "typical" poly-urethane molecule, since the structure is almost infinitely variable, according to the starting materials.

The term "polyurethane" is consequently something of a misnomer, since its structure is not represented by a polymerised "urethane monomer". The term is simply defined as a polymer that contains a significant number of urethane (-NHCOO-) linkages. These linkages may not necessarily repeat in any regular order, and are there to join larger blocks of other polymer chains, such as polyesters or polyethers.

Consumption of polyurethanes has grown steadily to reach about 3 million tonnes worldwide in 1979, the principal markets being upholstery foams for furniture, bedding and motor vehicles, and structural/insulation foams for construction and refrigeration purposes. The end-use markets in 1979 were divided up as shown in Table 3.1.

TABLE 3.1

Polyurethane end-use markets in 1979 [1]

Market	%
Furniture and bedding	40
Automotive	20
Building	11
Refrigeration	6
Textiles	6
Coatings	5
Footwear	4
Other	8

The types of material encompassed by polyurethanes may be conveniently divided into the categories of flexible foams, rigid foams, elastomers, coatings, and adhesives, of which foams constitute by far the largest outlet (85% of the total in 1979). Tin-based catalysts have had particular impact in the area of flexible foams, where they have enabled efficient production systems to be devised, but they are also used to varying extents in other polyurethane technologies.

3.1.2   The chemistry of polyurethanes

The urethane linkage is formed by the reaction between an isocyanate group and a hydroxyl group. The isocyanate group is very reactive and also combines with other functional groups that are present in the reaction mixture, so that several different reactions may be occurring simultaneously. The three important ones in foam production are chain propagation, gas formation, and cross-linking.

a)   Chain propagation. The addition of an isocyanate to an alcohol hydroxyl group produces a urethane:

-NCO + OH- $\longrightarrow$ -NHCOO-

Thus a reaction between multifunctional starting materials, such as a diisocyanate plus a diol, produces a polymerisation (more accurately, a polyaddition) to a polyurethane:

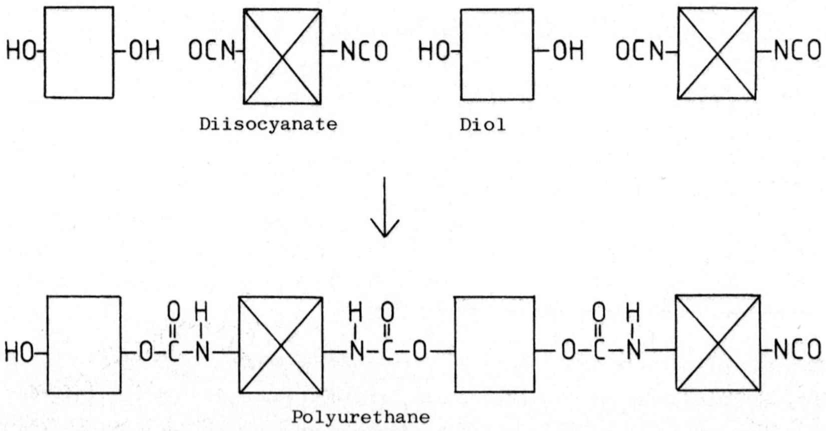

Diisocyanate   Diol

↓

Polyurethane

b) Gas formation. Isocyanates react with water to form an amine plus carbon dioxide gas, via an unstable intermediate:

$$-NCO + H_2O \longrightarrow -NHCOOH \longrightarrow -NH_2 + CO_2$$

Thus if water and a diol or polyol are present together to react with a di-isocyanate, gas formation will occur simultaneously with chain propagation, and the polymer will be blown out into a foam.

The amine from the above reaction can react further with isocyanate to form a urea linkage:

$$-NH_2 + OCN- \longrightarrow -NHCONH-$$

c) Cross-linkage. Additional side-reactions can occur during polymerisation that cause branching and cross-linking between chains. Thus isocyanate can also react with a urethane group itself to form an allophanate cross-link at higher

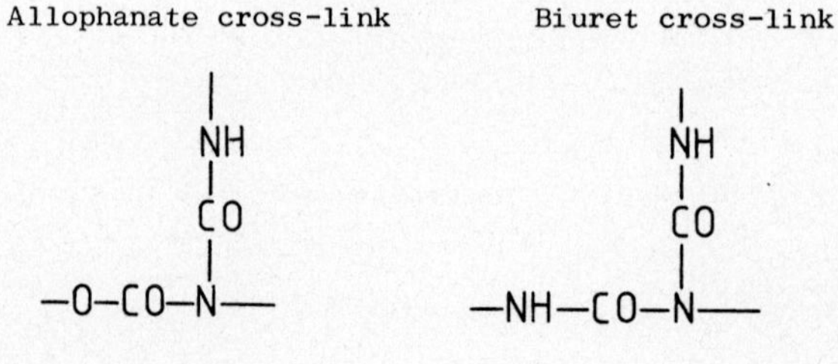

Allophanate cross-link   Biuret cross-link

temperatures, and it can react with a urea group to form a biuret cross-link.

Increased branching of the polymer chains leads to a reduction in properties such as solubility, flexibility, and elasticity, and an increase in tensile strength, hardness, and melting point. The polymers can thus be made into 3-dimensional networks by using an excess of diisocyanate over the stoichiometrically required amount, or by controlling the structure and functionality of the two principal constituents, the polyol resin and the isocyanate.

In foam production, the relative rates of the polymerisation and gas-producing reactions must be precisely controlled so that the setting of the polymer coincides with the maximum expansion of the foam. If the former is too fast, the material will set before it can fully expand, while if the latter reaction is too fast, the foam will expand quickly and collapse before the gas can be trapped. This balance is particularly critical for flexible foams, which require a completely open-cell structure for optimal resilience and flexibility of the material.

3.1.3 Historical development

As with several other plastics, work started on their development just before the Second World War, and was given extra impetus by the search for alternatives to valuable strategic materials such as natural rubber. The polyurethane reaction appears to have been first investigated by Dr. Otto Bayer of IG Farben Industries in Germany from 1937 onwards. He prepared polyurethane elastomers by reacting various polyols with tolylene diisocyanate, and attempted to develop flexible and rigid foams as well; some rigid foams were used in Germany to strengthen aircraft wings. During the War, work was also carried out in Britain by ICI, who developed a polyester rubber using diphenylmethane diisocyanate as a cross-linking agent; this material was used in barrage balloons. Du Pont were developing elastomers and adhesives in the U.S.A. at the same period.

In the late 1940s, Bayer AG and Hennecke AG in Germany developed a process for the continuous production of flexible foam, and ICI in Britain followed with their own system, working in conjunction with Viking Engineering. The chemicals supplied with these early processes included a polyester-type polyol, and an amine catalyst. Soon, however, polyether polyols were introduced, which enabled foams to be made from cheaper resins, and these polyether foams also matched the resilience properties of natural latex rubber foam more closely. Polyethers took over the major share of the flexible foam market, while polyesters are still used where special properties are required in the product. Subsequent developments in flexible foam block production include the "one-shot" process made possible by the advent of highly-efficient tin-based catalysts, as described later, and refinements in the production technique to minimise waste material that must be cut away from the foam block. Injection moulding techniques were also developed so that precise foam articles could be produced with no waste at all.

The earliest rigid foams used tolylene diisocyanate (TDI) in the formulation, but toxicity problems with these systems led to the development of rigid foams based on diphenylmethane diisocyanate (MDI) by ICI in the 1950s, together with specially designed dispensing equipment[2]. It was also discovered that the use of fluorochlorocarbons as blowing agents for rigid foam resulted in a material with the best thermal insulation properties yet found [3]. This was a major commercial breakthrough for rigid urethane foams, and led to the widespread production of the thin-walled refrigerator and other insulation applications.

With many companies entering the market, a wide range of sophisticated production machinery for polyurethanes has subsequently been developed, such as rigid foam laminating equipment for producing construction panels. Polyurethane elastomers, discovered early on in the 1940s, have started to find important markets relatively recently with injection moulded articles, particularly in the motor industry with items such as bumpers and other constructional parts.

Along with the developments in machinery, the major chemical companies have introduced a large number of proprietary formulations and blends for the manufacturer to buy "off the shelf" to convert into the desired products; depending on the manufacturing process, the total constituents for the foam formulation can be reduced to four, three, or even two components that are subsequently mixed together. As the ultimate, a "one component froth" that is used as a gap filler/insulator has also been developed; dispensed from a pressurised container, it expands and cures on reacting with moisture in the air or substrate.

3.1.4   Catalysts for foam production
The polymerisation and gas-forming reactions are both catalysed by certain classes of compounds, and catalysis is essential in order to obtain a reaction efficiency suitable for industrial production.

The first catalysts used for polyurethanes were tertiary amines such as N-alkyl morpholines, dialkylalkanolamines, and triethylenediamine; these catalyse the polymer and gas reactions about equally. However, the latter is much the faster reaction when uncatalysed, and the amine catalyst does not alter the balance between them. Consequently, the gas evolution is too fast for a controlled foam production. This is particularly true when polyethers are used, which are less reactive than polyesters.

This problem was originally overcome by using a two-stage process for foam production. In the first stage, the diisocyanate was reacted with a long-chain diol (say a polyether or polyester of molecular weight 2000-6000) in the presence of catalyst to form a long-chain diisocyanate pre-polymer. In the second stage, the pre-polymer was reacted with water plus catalyst to form the complete polymer plus carbon dioxide gas; the latter was efficiently trapped in the material since part-polymerisation had already been carried out.

It was evident that the efficient production of foam using a "one-shot" process whereby all the constituents are mixed together in one batch required catalysts that are much more effective for the polymerisation than the gas reaction. The most powerful such catalysts were found to be certain types of tin compound, which were identified and first used commercially in 1959. These include the organotins dibutyltin dilaurate and dibutyltin diacetate, and the tin(II) compounds tin(II) 2-ethylhexoate (usually called stannous octoate) and tin(II) oleate.

The effectiveness of some of these catalysts is illustrated in Table 3.2 which compares the effect of several compounds on the rate of a model reaction (phenyl isocyanate + 1-butanol).

TABLE 3.2
Comparison of catalytic efficiency of some organotins and amines [4]

Catalyst	Relative activity at 1.0 mole %
None	1
N-Methylmorpholine	4
Triethylamine	8
Tetramethyl-1,3-butanediamine	27
Triethylenediamine	120
Tetra-n-butyltin	160
Di-n-butyltin diacetate	56,000
Di-n-butyltin dilaurate	56,000
Di-n-butyltin dilaurylmercaptide	71,000

When a mixture of amine and tin catalysts is used (which is usually the case for flexible foams), the latter can be thought of as virtually controlling the polymerisation reaction, while the former mainly determines the gas reaction. A correctly balanced catalyst mixture is essential for obtaining the ideal flexible foam structure which has open cells for springiness but sufficient cross-

linked polymerisation for strength. It is also found that tin and amine catalysts have a synergistic effect when used together, so that the net catalytic
effect is greater than the sum of the individual parts.

The main application for tin catalysts (about 80-90% of the total) is in
flexible foam production, where the great majority of formulations (especially
for continuously-produced "slab" foam) employ some kind of tin catalyst. They
are also used, although to a lesser extent, for rigid foams, and minor quantities are used in the areas of elastomers and coatings.

The worldwide consumption of tin-based catalysts for polyurethanes in 1980
was estimated to be about 2,500 tonnes, of which easily the largest proportion
is stannous octoate. This compound, a clear pale-yellow slightly viscous liquid, is the standard catalyst employed for most flexible polyether foams. Originally dibutyltin dilaurate was used in this area, but it was found to promote
the high-temperature degradation of the foam; stannous octoate, however, has no
detrimental effect on foam stability. It is available under many different
trade names from chemical manufacturers, who include a small quantity of stabiliser with it to minimise oxidation by the air.

Where rigid foams employ a tin catalyst, dibutyltin dilaurate is usually the
one used. Its use in the U.S.A. in the early 1980s for this purpose amounts to
something over one hundred tonnes annually, and its corresponding consumption in
Europe is estimated to be of the same order. Dibutyltin dilaurate is also used
in certain specialised flexible foams, and finds some use in other polyurethane
formulations, including elastomers and coatings.

Mechanism. The catalytic mechanism of tin and other metal compounds in the
isocyanate-hydroxyl reaction has been the subject of a great deal of study over
the years, which has been reviewed extensively elsewhere [5]. The tin compound
is found to form complexes with both isocyanate and hydroxyl compounds, and the
catalytic effect is believed to come about by the formation of a ternary complex
of the tin with hydroxyl and isocyanate ligands [6]. This may be a multinuclear
bridge-type complex [7]. The isocyanate and hydroxyl groups are brought into
proximity in the complex, and subsequently react together; the exact mechanism
has not been clarified with absolute certainty, and some studies have indicated
that the mechanism is dependent on the precise reaction conditions, including
temperature, solvent, and type of ligand [8].

3.1.5   Foam formulation

The two major components of any polyurethane are the polyol and isocyanate, while the other essential components, including catalysts, water, surfactants, etc., form a relatively low proportion by weight. The types of polyurethane foam can be classified in various ways, the most convenient main categories being flexible and rigid foams, with an intermediate class termed semi-rigid. Within these categories, the foams can also be divided between slab foams which must be subsequently converted to the desired shapes, and moulded foams where the final shape is produced in situ by "reaction injection moulding" (RIM). Further sub-divisions can be made according to the chemical composition of the formulation, as described below.

(i) Polyols. These form the largest part by weight of any formulation, and a very large number of proprietary resins are available for the production of the great range of polyurethane-based materials. Flexible foams are divided into three types; a) polyether, b) polyester, and c) high resilience (HR), depending on the resin used.

Polyether polyols are the main type used in foam production, and they are manufactured by the polyalkoxylation of an alcohol with ethylene oxide and/or propylene oxide:

$$ROH + nCH_2-\underset{\underset{O}{\diagdown \diagup}}{\overset{\overset{R'}{|}}{CH}} \longrightarrow RO(CH_2\overset{\overset{R'}{|}}{C}HO)_n H \quad \text{Polyether}$$

The starting alcohol used determines the degree of branching of the polyether, and hence the degree of cross-linking produced in the final polyurethane. For flexible foams, polyethers based on divalent or trivalent alcohols are used (usually glycerol), having molecular weights in the range 2000-6000 for slab-stock production, or 3000-4000 for moulded foams. Rigid foams, on the other hand, employ multivalent highly branched polyethers based on compounds such as sorbitol, pentaerythritol, or sucrose; these usually have a molecular weight less than 1000. Semi-rigid foams usually employ polyether triols combined with low molecular weight glycols.

Whereas virtually all flexible upholstery foams are now based on polyethers, certain specialised uses require polyester foams. The former are more resistant to hydrolysis, but the latter are more resistant to organic solvents such as dry-cleaning fluids or oils, and combine this with good load-bearing properties.

Polyester foams (which now comprise less than 10% of all flexible slabstock production) are used in textile lamination, packaging, vehicle trim, etc. Polyester polyols are manufactured by the esterification of bi- or trifunctional alcohols (e.g. diethylene glycol) with bifunctional carboxylic acids (e.g. adipic acid), to a molecular weight of 2000-3000:

$$(n+1) HO(CH_2)_2OH \quad + \quad nHOOC(CH_2)_4COOH$$

$$\downarrow$$

$$HO(CH_2)_2 \left[OOC(CH_2)_4COO(CH_2)_2 \right]_n OH \quad \text{Polyester}$$

The presence of reactive primary hydroxyl groups in the polyester means that tin catalysts are not usually required for these foams.

The third and most recently-developed types of flexible foam are termed high-resilience, since they exhibit the highest degree of elasticity of any of the polyurethane foams, and also show minimal hysteresis in their load-indentation curves. Normal moulded polyether-based foams require a high-temperature curing cycle before the foam can be removed from the mould; however, high-resilience foams do not require this, and hence they are also referred to as "cold-cure" foams. They are produced from reactive starting materials, and various types of polyol resin have been developed in this sector. One type is a polyether triol of molecular weight 4500-6000 that has been "tipped" with ethylene oxide to produce a high content of primary hydroxyl groups. Another type are the "polymer-polyols" which are polyethers containing about 20% of an in situ copolymer of acrylonitrile and styrene. Reactive organic fillers (such as polyureas) may also be incorporated into polyethers, giving "PHD[*]-polyols". HR foams produced from tipped polyethers do not require a tin catalyst, but those made from polymer-polyols or polyurea-polyols usually employ dibutyltin dilaurate as part of the catalyst system. The use of tin catalysts in the various polyurethane foam systems is summarised later in Table 3.3.

[*] Trade name, Bayer AG.

(ii) Isocyanates. There are two principal isocyanate compounds that have
formed the basis of polyurethane technology since its inception, although over
the years a large number of modifications have been devised, together with some
new compounds for specific purposes.

Tolylene (or toluene) diisocyanate, TDI, is used for flexible foams of both
the moulded and slab type. Most flexible foam is made from "TDI 80", which contains 80% of the 2,4-isomer and 20% of the 2,6-isomer. Some specialised high
load-bearing foams use "TDI 65", containing 65% 2,4-isomer. The other main isocyanates are based on diphenylmethane-4,4'-diisocyanate, MDI. Pure MDI is used
in various modified forms for certain specialised products (e.g. "integral skin"
foams), but a polymeric form of MDI (PMDI) is a compound of major importance in
modern polyurethane technology; its general formula is shown in Fig. 3.1, and it
is a liquid mixture containing a proportion of pure MDI. It has the advantage
of a much lower vapour pressure than TDI, which reduces the possible health hazards during manufacture that can be caused by isocyanate vapours. Polymeric MDI
is widely used for rigid and also semi-rigid foams.

High-resilience flexible foams can use a wide range of specially-formulated
isocyanates. One common formulation is a mixture of 80% TDI plus 20% PMDI.
Other variants are produced by modifying TDI in order to produce certain adducts
or mixtures; thus it can be modified with urethane, TDI trimer, allophanate,
polyamine, etc. Similarly MDI is also modified with urethane or uretonimine for
special applications. The effect of these modifications is to increase the
functionality, i.e. the average number of isocyanate groups per molecule, and
thus increase the degree of cross-linking, consequently producing superior physical properties. Thus while a pure diisocyanate has a theoretical functionality
of 2.0, this can increase to 2.8 for PMDI and for anywhere between 2.0—2.8 for
the other variants used.

Certain other isocyanates have been introduced for applications such as polyurethane elastomers; these include naphthylene-1,5-diisocyanate (NDI), hexa-
methylene-1,6-diisocyanate (HDI) and isophorone diisocyanate (IPDI). Aliphatic
isocyanates such as HDI and IPDI are more light-stable than their aromatic analogues, and are used where a non-yellowing product is important, such as clear
or pigmented coatings.

It should also be noted that other reactions undergone by isocyanates are
used to produce other types of foam that are related to polyurethanes. For example, three molecules of a diisocyanate can form a trimer wherein three isocyanate groups combine to form an isocyanurate ring. Polyisocyanurate (PIR)

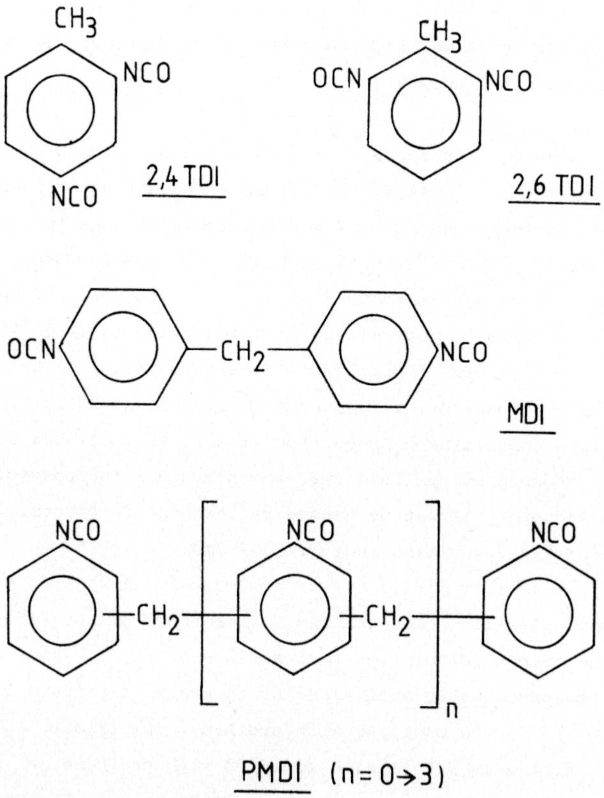

Fig. 3.1. The principal isocyanates used.

foams are rigid materials with superior heat- and fire-resistant properties, and are manufactured from a polyol plus a stoichiometric excess of isocyanate (compared to normal polyurethanes) using special catalysts; tin-based catalysts are not employed for these foams.

(iii) Catalysts. Stannous octoate and dibutyltin dilaurate are the principal tin-based catalysts used, in combination with amine catalysts, in many foam formulations. The main types of polyurethane foam are shown in Table 3.3 which shows the polyol and isocyanate components used, together with the type and typical concentrations of tin catalyst employed, the latter being in parts per hundred of polyol resin (phr).

Other tin-based catalysts are also commercially available for particular types of foam, including stannous oleate for flexible and dibutyltin diacetate

TABLE 3.3

The use of tin-based catalysts in polyurethane foams

Foam type	Polyol	Isocyanate	Tin catalyst
Flexible polyether	Polyether triols, MW 2000-6000 (slab) or 3000-4000 (moulded)	TDI	Stannous octoate, 0.1-0.4 phr (slab) or 0.1-0.2 phr (moulded)
Flexible polyester	Poly(diethylene glycol adipates), MW 2000-3000	TDI	Not usually used
Flexible high resilience	Modified polyethers, e.g. "tipped" triols MW 4500-6000, polymer-polyols, polyurea-polyols	TDI + PMDI, various modified TDIs and MDIs	Dibutyltin dilaurate used in some formulations, 0.01-0.15 phr
Semi-rigid	Polyether triols + glycols	PMDI usually	Dibutyltin dilaurate used in some formulations, up to 0.1 phr
Rigid	Multivalent polyethers, MW less than 1000	PMDI usually	Dibutyltin dilaurate used in some formulations, up to 0.1 phr

for rigid spray foams. In addition to being sold in the "pure" form, tin catalysts are also marketed in a diluted form, for example a 50% solution in a mineral oil; this provides for ease of handling and greater accuracy in controlling
small quantities of catalyst.

(iv) Water. Water is present to react with isocyanate and produce carbon dioxide gas, which blows out the reacting mixture into a foam. Thus the more water present, the lower the density of foam produced. The chemical structure of
the foam will also be altered, however, since the amine produced by the water-
isocyanate reaction can react with more isocyanate to produce first a urea and
then a biuret structure, thus increasing the degree of cross-linking. Water is
used up to a level of about 5 phr. Foams can be purely water-blown, but it is
common to have an auxiliary blowing agent as well.

(v) Blowing agent. The common blowing agent used is monofluorotrichlorometh-
ane, $CFCl_3$ (also called Refrigerant 11), an inert liquid of boiling point 24°C.
The polymerisation reaction is highly exothermic and vaporises the blowing
agent, thus supplementing the foam expansion and producing a lower-density foam,
while not affecting its chemical structure. Flexible foams have a completely
open-cell structure, and thus no blowing agent is trapped in the foam; however,
rigid foams may have a completely closed-cell structure, trapping the R11 vapour. Since the latter has a very low thermal conductivity, the effect is to
produce an extremely efficient insulation material. The blowing agent may comprise up to about 15 phr in the formulation, and foams may be purely R11-blown,
with no water used.

Various other compounds have been investigated as blowing agents in recent
years owing to possible legislation on the emission of fluorochlorocarbons into
the atmosphere. Dichloromethane, CH_2Cl_2, is now also used in various formulations.

(vi) Surfactant. Silicone surfactants or stabilisers are essential to homog-
enise the reaction mixture, reduce the surface tension, and promote a stable and
even growth of gas bubbles. These are often polysiloxane-polyether copolymers,
used at a level of about 1 phr.

In addition to the above constituents, there are various others that may be
included, for example pigments, anti-ageing agents, various types of filler, and
flame-retardants. A light organic material such as a flexible foam burns easily, and the incorporation of a flame retardant is usually essential; compounds
such as chlorinated phosphate esters are common, and may constitute up to 10% by

weight of the formulation. High-resilience foams have improved fire-retardancy over the normal flexible type.

3.1.6   Production processes

The successful mass-production of high-quality polyurethane foams, either as a continuous slab or moulded articles, has required the development of very precise technology over a number of years, and indeed this is still continuing today, along with continuing developments on the chemistry side to match the demands of the machinery. Basically, any mass-production system comprises the following:

a)   The foam constituents are stored in tanks where they are conditioned to the right temperature.

b)   The constituents are pumped at precisely controlled rates to a mixing head, where they are thoroughly mixed together.

c)   The reacting mixture is laid down on a moving conveyor for continuous slab production, or is deposited into moulds.

d)   After suitable conditioning, the foam slab is converted into the desired forms, or the articles are removed from the moulds.

Conditioning and maintaining the foam constituents at the correct temperature (usually 25°-30°C) is essential since any variation can affect not only their reactivity, but also their viscosity. The storage tanks are usually fitted with water jackets, and heat exchangers in the circulation system also provide temperature control; the components in the storage tanks are generally mechanically agitated to ensure homogeneity.

Each constituent of the foam formulation can be pumped individually to the mixing head, but it is common for pre-blending to be done so that the total number of feed-streams is reduced to four, three, or two; two-component systems are often employed for rigid foams and reaction injection moulding. Pre-blending of the catalysts, blowing agent, surfactant, etc., is normally done with the polyol, and this is carried out either at the manufacturing site, or beforehand by the raw material supplier.

Foam production machines may be of two basic types, either low pressure or high pressure. In the former, the constituents are pumped under low pressure into the mixing head, where vigorous mechanical agitation ensures that thorough

mixing takes place as the material is dispensed from the head. In the high
pressure system, the constituent streams are made to impinge into each other under high pressure, and no mechanical agitation is needed. Low pressure systems
are often used for continuous flexible foam production, while high pressure systems are used in reaction injection moulding since they ensure an accurate and
reproducible shot with mixing heads that are self-cleaning. In operation, the
various components of the foam are continuously pumped and recirculated back to
the storage tanks prior to pouring, usually via the mixing head itself, although
some types of high pressure system use a recirculation loop that by-passes it.
To commence foam pouring, the valve in the mixing head is opened (or alternatively, the by-pass loop is closed off) allowing the components to mix in the
desired ratios in the chamber before pouring out of the nozzle.

(i)   Continuous flexible slabstock. Production is carried out on long convey-
ors, the first of which receives the reacting mixture from the mixing head.
This primary metal-slat conveyor is lined on three sides by a U-shaped trough of
paper that is continuously drawn from large storage reels at one end. In one
common process, the mixing head oscillates steadily across the width of the conveyor with the dispensing nozzle just above the surface in order to prevent
splash and air entrapment. The primary conveyor is fixed at an angle of between
$2°$ and $5°$, so that the initially-fluid mixture flows away from the nozzle and

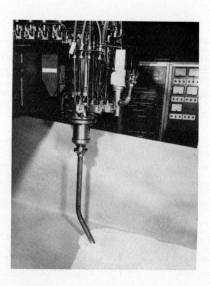

Fig. 3.2. The reaction mixture is laid down in a paper trough from an
oscillating mixing head.
Photograph courtesy: Dunlop Ltd., Dunlopillo Division, High Wycombe, U.K.

into (but not over) the just-foaming polymer; the point at which the mixture just starts to foam and expand is the "cream line". By the end of the primary conveyor, the foam has risen to its full height, and passes onto a secondary horizontal conveyor where it is cut into sections by a vertical travelling band-saw; the foam production section is enclosed in a fume tunnel to extract the gases evolved. The exothermic nature of the reaction may cause the temperature at the centre of the slab to reach 170°C, and no extra heat needs to be applied to cure it. However, the foam blocks must be stored away from each other for about 12 hours until they have cured completely and cooled down; they can then be stacked together and subsequently processed.

Modern plant machinery may have outputs of up to 500 kg/min, producing a foam slab (or "bun") up to 1.2 m high and 2.2 m wide, with the length only being lim-ited by storage space. The height of the foam slab is determined by three fac-tors, a) the polymer formulation, b) the nozzle output, and c) the conveyor speed, while on certain types of plant it is possible to vary the width of the slab without stopping production.

The foam blocks must be subsequently trimmed of unusable material, such as the denser outer skin. However, the major potential cause of waste is the ten-dency of the foam to rise into a loaf-shape with a pronounced top curvature, caused by frictional drag between the foam and the side-walls. This curved part

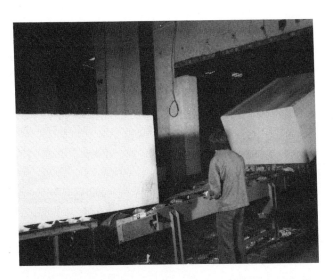

Fig. 3.3. Fully-risen foam (here produced by the Maxfoam process) is cut into blocks and transferred to the cooling/curing section.
Courtesy: Draka Foam Ltd., Glossop, U.K.

72

must be trimmed away before conversion, and thus it is desirable to produce a foam slab with as flat a top as possible. Various techniques have been devised to achieve this, the main ones being:

a)   The Draka Square Block * process. In this technique (in commercial operation since 1967), a polyethylene film is fed in between the rising foam and the paper side-walls, and it is pulled upward in an ascending profile that matches that of the rising foam. This reduces the frictional drag at the side-walls, producing an almost flat top surface [9].

b)   The Planiblock * process. This method, developed in the early 1970s, utilises a paper web that is continuously fed onto the surface of the rising foam, and is kept in position by a series of counter-balanced platens. The pressure exerted prevents the doming effect, and yields of up to 94% of prime foam are claimed for this process, which is virtually a form of moulding [10].

c)   The Maxfoam * process. This rather different approach was also developed in the early 1970s [11]. Here, the foam constituents enter a stationary mixing head, from which the mixture is fed via two pipes into the bottom of a metal trough. As the mixture starts to foam, it rises and overflows the trough, and is deposited on a moving belt of paper. This carries the foam down an inclined

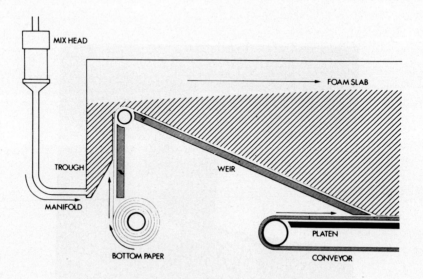

Fig. 3.4. The Maxfoam process.

* Trade names throughout.

pour-plate or weir, which is arranged so that the foam's expansion in this section takes place mainly in a downward direction relative to the side-walls, i.e. the rate of rise of the foam is virtually matched by the rate of drop along the incline. This produces a flat-topped block, with a claimed yield of 93-94% of prime foam.

In addition to rectangular foam blocks, different cross-sections can be produced by shaping the paper web accordingly using special machinery; thus cylindrical blocks, for example, are also manufactured.

The final foam blocks obtained are fabricated or converted using a wide range of techniques and equipment. Simple or contoured cutting and shaping is done by a vibrating blade or hot wire, and an almost unlimited variety of parts can be fabricated using these and other techniques such as stamping, convoluting, skiving, and buffing.

(ii)   Reaction injection moulding. Moulding techniques are more efficient and economical where the continuous mass-production of a precisely-contoured shape is required, for example car seating. In addition, there are the moulding processes where the polyurethane material is injected into its final location and is not removed, for example the rigid cavity insulation foams for buildings. The RIM techniques have made great strides in recent years with the development of increasingly sophisticated equipment and formulations.

There are many types of moulding operation, encompassing flexible and rigid foams; in a typical set up for flexible polyether moulded parts, the production line is in the form of a loop with a stationary pour station. The mixing head dispenses a shot of polymer mixture into a sequence of empty metal moulds which have been preheated to 35-40°C and lined with a wax releasing agent; they may also contain an appropriate insert such as a seat frame. The closed moulds then pass through a convection oven in the temperature range 150°-250°C in order to cure the foam, and this process may take about 10-15 minutes. On emerging from the oven, the moulds are opened and the foam part demoulded; the moulds are then re-conditioned by cleaning, applying new release agent and cooling to the right temperature for the start of the cycle. The whole moulding cycle may last about 20 minutes, depending on the size of the part.

The use of a high-resilience "cold cure" foam means that the curing oven is not required, although the mould temperature required is slightly higher at 45°-55°C. The curing time required within the mould is often in the range 4-8 minutes, but the most recent systems permit demoulding in as little as 2 minutes.

Thus there is a saving here in time and energy costs, although HR formulations tend to be more costly.

Other mass-production moulding systems have been devised, depending on the size of the moulds and the output required. For example, there can be stationary moulds with a mobile mixing head; alternatively, the feed tanks can supply a series of up to a dozen mixing heads that are fixed directly to stationary moulds. Reaction injection moulding has been a growth area not only for foams, but also for polyurethane elastomers, which are being increasingly employed in vehicle parts such as bumpers; the applications of moulded polyurethanes have been further extended by the development of techniques for the production of filled or reinforced mouldings and for self-skinned foams, as described below.

Reinforced reaction injection moulding (RRIM): The properties of polyurethanes can be modified by incorporating a variety of fillers or reinforcements, and materials such as silica, rubber, and glass fibre have been used. Glass-fibre reinforcement in particular has shown great potential in producing high strength rigid foams which have the additional benefits of low density and good surface properties. Such materials are being increasingly incorporated into motor vehicles in order to reduce weight and lower fuel consumption.

Integral-skin foams: In use, polyurethane foams are often covered by an outer skin of another material such as a fabric or PVC. However, the bond between the foam core and the outer skin may rupture in service if subjected to great stresses, and techniques have been devised to produce an integral moulding in one step which has a cellular core and a non-porous outer skin, both in polyurethane. Such "integral skin" (or "self-skinning") foams result from a careful control of the reaction conditions and formulation; the high temperature at the mould core vaporises the blowing agent present and produces a pressure against the metal mould wall which is at a lower temperature. The combination of pressure and lower temperature condenses the blowing agent in this region, forming a non-cellular skin. Integral-skin mouldings are produced for flexible and rigid foams, and also microcellular elastomers.

(iii)   Rigid foam techniques. Unlike flexible foams, which have a completely open-cell structure, rigid foams usually have a closed-cell structure, although open-cell types are used for specific applications. Among the various ways that rigid foam can be processed are those techniques already mentioned, namely continuous slabstock production and reaction injection moulding.

Continuous slabstock production is the most economical method for long production runs, and the process is similar to that described for flexible foam, with the reaction mixture being laid down on a moving conveyor belt. A Planiblock system can be used to ensure a rectangular final block, which is subsequently sawn into boards suitable for construction, etc. A simple discontinuous method can also be applied for small-scale production; here the raw materials are simply mixed together by hand and poured into a suitable mould such as a rectangular box, or alternatively the reaction mixture can be dispensed from a machine.

As mentioned earlier, rigid polyurethane foam containing R11 vapour in its closed cells is the most efficient thermal insulation material known, with a thermal conductivity of the order of 0.02 W/mK or less; furthermore, the vapour has a very low permeability through the cell-walls, and hence the insulation properties are retained, particularly when a foam panel is combined with an impermeable facing material. The production of rigid polyurethane composite boards with a range of other laminate materials has become an important sector of the industry, and various types of machinery have been developed for the continuous production of "sandwich panels" with such facings as paper laminates, felt, aluminium foil, glass mat, steel sheet, plasterboard, etc.

The continuous production of sandwich panels with fairly flexible web facings is carried out by a double conveyor technique, whereby the polyurethane reaction mixture is laid down by a mixing head onto a continuous belt of the facing material. As the mixture expands, it comes into contact with an upper parallel belt of material that is fed in from another roll. The outer webs are held rigidly by steel platens in order to withstand the pressure of the reaction, and after the foam has cured, the continuous panel is automatically cut into individual boards which are stacked; an output of 10 m/min may be typically attained, producing panels up to 20 cm thick. This process takes advantage of several properties of rigid foam systems, including the excellent adhesion to most surfaces, good flow properties so that it fills even complex profiles, and fast curing. The last property is aided by organotin catalysts, which are particularly beneficial in this type of system. Developments in machinery have led to the continuous production of panels with a rigid facing on one side, for example asbestos cement. In the inverse lamination technique, the discrete rigid sheets are fed onto the conveyor as the lower facing, while the foam mixture is laid down on the flexible material that will become the upper facing. The foam expands and partially cures, while the backing web moves around a semicircle of platens that inverts it and brings the foam into contact with the pre-heated rigid sheets; the two layers are pressed together in a double conveyor arrangement, after which the bottom carrier paper is rewound and the board is cut into

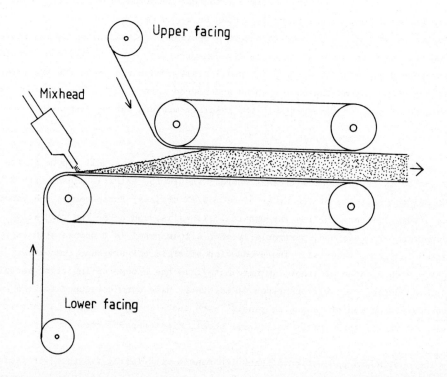

Fig. 3.5. The double conveyor technique for producing panels with a rigid foam core.

the individual panels. A further development of this system is the duplex lamination method, where both the upper and lower facing materials are given a layer of foam, using an extra mixing head, before the two foam layers are brought together to form a single sandwich element.

It is also possible to profile the facing materials before they are filled with rigid foam. This is done with steel sheet, where both the upper and lower facings can be fed from stock coils through roll formers that impart the desired profiles into the sheets; the filled panel is then produced by the standard double-conveyor method. While the steel sheets naturally add strength to the polyurethane core, the use of profiled facings gives extra high rigidity that enables the boards to be used directly as structural elements such as roofs or wall sections.

In addition to the continuous production of sandwich panels, discontinuous methods are used where only a low volume of production is required. This method is suitable for thicker panels and also for those with rigid facings on each side. Here, the facing materials are placed in a press, and the reaction mixture is dispensed into the cavity between them; the panel is removed after foaming and curing is completed.

Rigid foams are also applied by on-site foaming at the particular locations where it is required, using pouring or spraying techniques. This is a convenient and cost-effective way of applying insulation foams, for example, since the transportation and storage of the liquid chemicals required (normally comprising two components) is easier than handling the bulky final material. The spraying of insulation foam is carried out with high-pressure equipment, and a very reactive formulation is required so that the foam quickly cures and cross-links, and does not drip off vertical walls or ceilings; organotin catalysts find application in this area.

3.1.7 Applications

(i) Flexible foams. The properties of flexible foams such as density, load-bearing, and resilience can be tailored to suit a range of end-uses. The durability of a foam depends to a large extent on the density, and while flexible foams are produced over a range of densities from 15 to 50 kg/m^3, those intended for hard-wear use, such as seat cushions, require a minimum density of about 35 kg/m^3. Furniture is the largest single market for flexible foam, with approximately 40% of the total. By combining foam grades with different densities, load-bearing properties and resilience, a high degree of cushioning comfort can be obtained; there is increasing use here, as in other upholstery applications, of high-resilience cold-cure foams to give greater quality of comfort and a more efficient manufacturing operation.

Bedding utilises about 25% of the flexible foam market, and mattresses can either have a 100% polyurethane core or a combination of springs with foam "topper pads". Flexible slabstock has several advantages for this application, including its lightness, high air permeability, and ease of cleaning and sterilisation, which is important in institutions such as hospitals.

Approximately 20% of flexible foam is used in transport applications, where it fulfils the need for high comfort and improved safety aspects. Car seating may either be produced from slab foams, or moulded around a frame directly; the design of such seating is arranged to give maximum damping of the vibrations created by the vehicle. In addition, foams of different densities are used in arm and head rests, crash pads, door trim, sun visors, and as acoustic insulation generally in the vehicle. For safety padding in various transportation

Fig. 3.6. The motor industry employs a wide range of flexible and semi-flexible foams in seating, head- and arm-rests, door trim, and safety padding.

applications, the class of foam termed semi-flexible (or semi-rigid) is often used; this has over 90% of open cells in its structure, and a density of between 100 and 200 kg/m^3. This results in a higher load-bearing capacity than flexible foam, together with lower elasticity and a slower compression recovery rate, making this type of foam ideal for shock absorbing. It is normally moulded into outer skins of PVC or ABS for use in protective panels, consoles, armrests, etc. Flexible and semi-flexible foams are used in the aircraft industry in analogous applications to those in the motor industry.

The lamination of textiles represents a significant outlet for flexible foam (6%), where the polyester type is generally used because of its higher mechanical strength and better resistance to oxidation and organic solvents. The resulting materials are used in windcheaters, coats, sleeping bags, etc. A growing use for urethanes is as a carpet underlay, where various techniques have been developed for applying the foam as a backing.

Packaging is a well-established use for flexible foam (approx. 2% of total), and a wide range of fragile goods, or goods that must be protected from excessive shocks, are packed in foams which are often fabricated to the precise shape required. The grade of foam used depends on the nature of the article to be packed, and a combination of light and dense foams can be used to give maximum protection.

There are numerous other small markets for flexible foams, which nevertheless add up to a sizeable usage. These include household sponges, sealing tapes, sound insulation, thermal insulation for hot water tanks and pipes, and air filters.

(ii)   Rigid foams. The main advantages of these materials have already been mentioned, namely the unsurpassed insulation properties of those blown with R11 (approximately twice as efficient as polystyrene, cork, mineral wool, or fibreboard), the high strength to weight ratio, and the excellent adhesion to many types of material. Rigid foams also have good heat- and fire-resistant properties, and can normally withstand temperatures of $-200^{\circ}C$ to $+130^{\circ}C$ in use; when subjected to very high temperatures, there is a tendency to char rather than ignite, hence they do not present a fire hazard.

Rigid foams can be produced with densities normally ranging from 10 to 300 kg/m^3 by varying the formulation, although most insulation foams are between 20 and 40 kg/m^3. As a combined insulation/structural material, increasing use is found in the building industry, particularly with the growing awareness of the need to conserve energy, and the growing requirements for readily-erected construction units to help solve some of the world's housing problems. Thus polyurethane board, which can be laminated with a wide range of materials, is employed in exterior wall elements, roof sections, interior walls and doors. For added strength, various fillers can also be used in the core, and foam cores enclosing a wire mesh framework have been produced as another type of lightweight yet strong structural panel. On-site pouring or spraying of rigid foam is used, as mentioned earlier, as an efficient and cost-effective method of applying insulation to wall-cavities or areas of roofing. The spray technique is particularly useful for applying insulation to awkward shapes and contours, such as pipes.

Polyurethane insulation is also widely employed in refrigerators and freezers, industrial tanks, and refrigerated vehicles. The thickness of foam required for a given degree of insulation is much less than that required from alternative materials; this results in a gain in capacity for a refrigerated container with a given set of exterior dimensions.

Closed-cell rigid foam is an excellent flotation material, and is used as such in various marine applications such as buoys, small pleasure boats, and barges. It has been found to be particularly effective for renovating old leaking barges, which can be subsequently rendered leak-proof and virtually unsinkable by injecting foam into the compartments above the hull.

Self-skinned rigid foams (with densities up to 800 kg/m^3) are produced for structural applications where they compete with more traditional materials such as wood, metals, glass-reinforced plastics, and solid or foamed thermoplastics. Thus mouldings are utilised in furniture, audio equipment, housings for office machines, and various constructional parts. Rigid polyurethane foams which are thermoformable have also been developed, and these, in combination with surface decorative layers, have found use as interior parts in motor vehicles.

(iii)   Rebonded composites. The fabrication of slab foam into various finished articles naturally leads to the generation of some scrap foam from parts that are cut away, and this scrap is usually made use of by granulating it into small crumbs and rebonding the crumbs together. This can be done with the use of suitable adhesives, and by moulding the composite under pressure, a fairly dense foam product with good load-bearing properties can be obtained, suitable for packaging or safety padding. An alternative method of producing composites is the "double foaming" process, first developed in the 1950s. Here, the granulated scrap foam is mixed with further reactants, namely polyol, isocyanates, catalysts, surfactant, etc., and the mixture is reacted together in a mould. Other conditions may be applied as appropriate, for example heat or steam. In this process, the granulated foam becomes highly compressed and is held this way by the new foam that forms around it. This is claimed to result in improved mechanical properties and better long-term stability for the foam.

One British company has developed a range of composite materials based on the double foaming technique, using stannous octoate in the catalyst system for the final polymerisation. One series of foams is produced with densities in the range 60-230 kg/m^3, and these are employed for cushioning and padding in furnishings and motor trim, as well as in traditional packaging outlets. Developments from these materials have led to more highly compressed composites, having densities in the range 200-600 kg/m^3. These have been found to be suitable for footwear inserts and mouldings, where they are claimed to be durable yet lighter than traditional materials used. The most recent development is an even more compressed polyurethane composite that resembles a non-porous, flexible, rubber-like material, with a density of 930 kg/m^3. This is used as a soling material for footwear, where it is claimed to be tough, non-slip, and will not mark floors. The double foaming method is also utilised to produce composites of polyurethane with other materials; a polurethane/rubber composite, for example, is manufactured as a durable, non-slip safety flooring for industrial and recreational applications.

Fig. 3.7. Highly compressed composites produced by double foaming are suitable as footwear inserts.
Photograph courtesy: General Foam Products Ltd., North Shields, U.K.

Fig. 3.8. A very highly compressed composite used as a soling material.
Photograph courtesy: General Foam Products Ltd., North Shields, U.K.

(iv)   Elastomers. Polyurethane elastomers are solid rubber-like materials that encompass a range of chemical structures, application methods, and uses. They may be produced from one- or two-component systems, and can be applied by techniques such as casting, moulding, spraying, brushing, rolling, etc. Microcellular elastomers contain a very fine dispersion of gas cells in the matrix that reduce the overall density, making the material particularly suitable for shoe soling and automotive applications.

Dibutyltin dilaurate is used as a catalyst in some elastomers, including certain casting and moisture-cured coating formulations. Examples of applications include car steering wheels, dashboard sections, rollers, abrasion-resistant linings, surface primers, sealants, and weatherproof membranes. Many coating systems make use of moisture-curing polyurethane intermediates as the starting material; these are based on liquid pre-polymers which typically consist of long-chain polyether- or polyester-based polyurethane molecules having terminal isocyanate groups. They are manufactured and compounded with other additives in anhydrous conditions, but after applying to a desired surface, a reaction with atmospheric moisture occurs that causes the material to cure to a tough, chemically resistant, elastomeric coating. (Alternative curing systems involve the addition of a diol or amine component to the pre-polymer).

In one-component moisture-cured systems, the initial reaction is isocyanate-water, which produces terminal amine groups on the pre-polymer, via the unstable carbamate; the amine groups then rapidly react with other isocyanate end-groups to produce urea cross-linkages (see section 3.1.2). The formation of the urea cross-links results in the curing of the elastomer. Note that the function of an organotin compound in this type of system is to catalyse the isocyanate-water reaction, not the more usual isocyanate-polyol reaction; for this purpose it is rather less efficient than some amine catalysts, but it is often undesirable to have too reactive a system here, otherwise there can be practical problems in application.

One British company markets a series of urethane pre-polymers suitable for use as primers, sealants, and roofing membranes. In the production of flexible, weatherproof roofings, a 1:1 blend of pre-polymer with special types of tar or pitch is often used. Other constituents to be added include a solvent diluent, defoamer (to aid release of carbon dioxide evolved), moisture scavenger (to remove inherent moisture before application), catalyst, and possibly other thickeners; dibutyltin dilaurate is employed at 0.25-0.5% by weight as catalyst. This pitch-based roofing system is claimed to be relatively inexpensive with low moisture permeability and a high resistance to water and UV. Alternative roofing systems can employ aluminium instead of pitch in the formulation.

Another advantage of this type of applied elastomeric coating is the fact
that it is seamless, with no potential points of weakness. In addition to
roofs, polyurethane-based coatings are also used to produce attractive, hard
wearing, seamless floorings by compounding the pre-polymers with suitable pigments and other additives.

3.2   SILICONES

3.2.1   Introduction

The term "silicone" is used for a very wide range of polymeric products based
on the inorganic siloxane backbone, $(-Si-O-)_n$, the other valencies on the silicon atom usually being filled by organic groups. The silicon-oxygen bond is
particularly strong and confers a high stability and chemical inertness to these
polymers. Silicones may be produced as resins, oils, gums, elastomers, or
pastes, and their wide range of uses reflects their diversity of physical form
and chemical structure. For example, they have already been mentioned in this
chapter as stabilisers for polyurethane foams; the compounds employed here are
relatively non-reactive surfactant oils that can be used to control or eliminate
foam formation in many processes. Such inert liquid silicones also form the
basis of damping and dielectric fluids, and various types of lubricant and release agent.

The scope of silicone technology was considerably extended with the introduction of silicone compounds that are fluid enough to be easily poured or spread,
and then subsequently cure to a tough elastomeric solid at room temperature.
While silicone rubbers that are cured, or vulcanised, by the action of heat are
also known, the development of elastomers that cure at room temperature introduces obvious benefits for ease of utilisation. Organotin compounds are one of
the principal types of catalyst used in these room-temperature-vulcanising (RTV)
silicones.

The main attributes of the various silicone products are their retention of
physical characteristics over a wide range of temperature, typically $-50°C$ to
$+250°C$, although this range can be extended. These characteristics include
flexibility (for the elastomers), electrical and thermal insulation properties,
dimensional stability, chemical inertness, and high water repellency. In addition, they are physiologically non-toxic and easy to handle. This combination
of properties has led to their extensive use as protective coatings, sealants,
moulding compounds, etc., which are discussed later.

3.2.2   Silicone chemistry

Most silicone polymers are based on polyorganosiloxanes, where the group bonded to the silicon is usually hydrogen or a simple organic group such as methyl or phenyl. Methyl groups impart high water-repellency, hardness, and good release properties, while phenyl groups impart good heat- and oxidation-resistance, good compatibility with organic products, and abrasion resistance.

The silicones range from free-flowing liquids to viscous gums, depending on the degree of polymerisation. In order to develop elastomeric properties, a curing or vulcanisation process has to take place so that the linear polymer chains cross-link to form a network, and there are two principal types of curing reaction, namely condensation and addition. In the condensation reaction, poly-dimethylsiloxanes are used which have terminal hydroxyl groups on the chains; these react with a cross-linking agent, namely a tetraalkoxysilane, in the presence of a catalyst. Four chains thus become linked via new siloxane bonds, and an alcohol is evolved as a by-product:

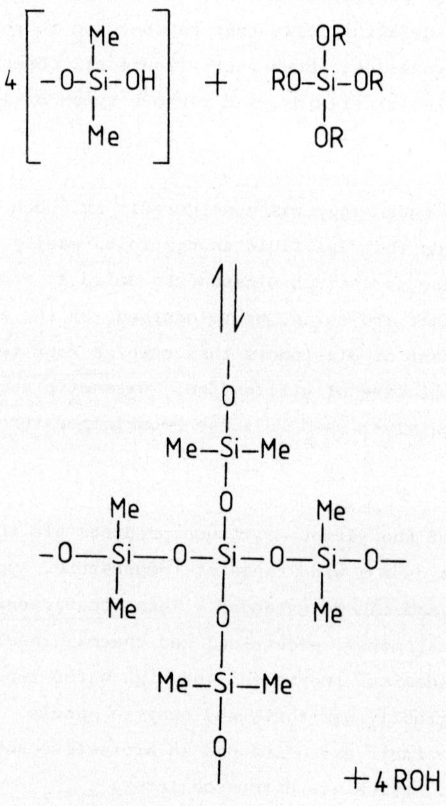

$$
4 \left[\begin{array}{c} Me \\ | \\ -O-Si-OH \\ | \\ Me \end{array} \right] \quad + \quad \begin{array}{c} OR \\ | \\ RO-Si-OR \\ | \\ OR \end{array}
$$

$$
\big\Updownarrow
$$

$$
\begin{array}{c}
| \\
O \\
| \\
Me-Si-Me \\
| \\
Me \quad O \quad Me \\
| \quad\quad | \quad\quad | \\
-O-Si-O-Si-O-Si-O- \\
| \quad\quad | \quad\quad | \\
Me \quad O \quad Me \\
| \\
Me-Si-Me \\
| \\
O \\
| \quad\quad\quad +4\,ROH
\end{array}
$$

Fig. 3.9. The condensation curing reaction.

The silicone precursor normally contains fillers or other additives to control the consistency of the uncured product or to control the mechanical and physical properties of the final elastomer. The evolution of a by-product in the above reaction means that the silicone elastomer shrinks slightly while it is curing. It will also be noted that the condensation reaction is reversible, and in fact reversion can occur at elevated temperatures; this means that condensation-cured elastomers must not be subjected to high temperatures until they are fully cured and the alcohol by-product has dissipated.

Addition vulcanisation utilises a polydialkylsiloxane that has vinyl termination groups, while the other component of the system is a polyalkylhydrogensiloxane. The silicon-hydrogen group of the latter compound adds to the vinyl double bond in the presence of a catalyst, forming a $-CH_2-CH_2-$ cross-link:

$$-O-\underset{\underset{R}{|}}{\overset{\overset{R}{|}}{Si}}-CH=CH_2 \; + \; H-\underset{\underset{O}{|}}{\overset{\overset{O}{|}}{Si}}-R' \; \rightarrow \; -O-\underset{\underset{R}{|}}{\overset{\overset{R}{|}}{Si}}-CH_2-CH_2-\underset{\underset{O}{|}}{\overset{\overset{O}{|}}{Si}}-R'$$

There is no by-product to this reaction, which is not reversible and is accelerated by heat.

Most condensation-cured RTV elastomers employ a tin-based catalyst, while the addition-cured type normally require a platinum compound; the remainder of this section will thus concentrate on the former, which are the most widely-used type. The catalyst normally used in condensation-cured elastomers is dibutyltin dilaurate, although other compounds are also employed, including dioctyltin dilaurate, dibutyltin diacetate, dibutyltin di-2-ethylhexoate, and tin(II) 2-ethylhexoate (stannous octoate). The catalyst may be incorporated at a level of 0.1 to 1.0%, although in two-part systems it is often diluted with a silicone (for example, the cross-linking agent) or a solvent, and it may also be bulked with a filler; this makes the addition less critical when mixing the components together.

3.2.3   Catalytic mechanism

The mechanism of the catalysis of condensation-cured RTV elastomers by organotin compounds has been studied by several groups of workers, and as often occurs in these situations, the conclusions do not always agree. However, it has been found that when dialkyltin dicarboxylates are present in the reaction mixture, no vulcanisation takes place under anhydrous conditions [12]. Since other protic substances present in the mixture do not start the vulcanisation process, it is deduced that water is not acting here as a protic co-catalyst, but that a hydrolysis product of the dialkyltin dicarboxylate is formed as the actual catalyst, namely the hydroxycarboxylate [13]:

$$R_2Sn(OCOR')_2 + H_2O \rightleftharpoons R_2Sn(OCOR')OH + R'COOH \tag{i}$$

This is supported by the fact that the addition of a carboxylic acid to the mixture (which shifts the equilibrium to the left) retards the cure [12]. It is proposed that the organotin hydroxycarboxylate reacts with the ethoxysilane cross-linking agent to form an organotin silanolate [12]:

$$R_2Sn(OCOR')OH + Si(OC_2H_5)_4 \rightleftharpoons R_2Sn(OCOR')OSi(OC_2H_5)_3 + C_2H_5OH \tag{ii}$$

The silanolate then reacts with polydimethylsiloxane to yield a functionally-substituted silicone compound, with the catalyst regenerated [13]:

$$H[OSi(CH_3)_2]_nOH + R_2Sn(OCOR')OSi(OC_2H_5)_3 \rightleftharpoons R_2Sn(OCOR')OH$$

$$+ H[OSi(CH_3)_2]_nOSi(OC_2H_5)_3 \tag{iii}$$

The substituted silicone compound can subsequently undergo further reactions with the catalyst and polydimethylsiloxane until all its accessible ethoxy groups become replaced by polydimethylsiloxane chains; thus up to four of the latter become bonded together via the cross-linking agent. Although many compounds are able to catalyse the reaction between silanols and alkoxysilanes, the organotin dicarboxylates are stated to be particularly useful because of their selectivity, that is, they do not catalyse undesirable side-reactions to any great extent at ambient temperature [13].

3.2.4 Applications

RTV silicone elastomers are available as either one-pack or two-pack systems. The two-pack systems were the first to be developed, and as the name implies, they comprise two separate compounds that are stable until they are mixed together, when vulcanisation begins. One component consists of the catalyst, often blended with cross-linking agent; the main component comprises one or more polysiloxane polymers, which may also contain various fillers (e.g. silica, calcium carbonate, iron oxide, china clay, etc.) and pigment, and also cross-linking agent if this is not contained in the catalyst component. Two-pack systems are available with a range of viscosities from pourable liquids to pastes, and the user can choose the most suitable type for a particular application. The curing properties can also be chosen as optimal according to use, and the following stages of cure are usually referred to for a given formulation:

a)   The pot life, which is the time from the addition of curing agent until the mixture ceases to flow; this is effectively the period that the material can be worked.

b)   The tack-free time, which is the time from the addition of curing agent until the material loses all surface tack.

c)   The cure time, which is the time from curing agent addition until a fully cured elastomer is obtained. This is often of the order of 24 hours for many formulations at room temperature.

The curing process for two-pack elastomers can be quickened by increasing the concentration of catalyst present, and also by raising the temperature up to about 70°C maximum if desired. Since the catalyst is thoroughly dispersed throughout the mixture, two-pack elastomers can cure in deep sections, which is essential for many applications such as potting. Adhesion to other surfaces such as metal, glass, plastics, etc., is generally poor, but very good adhesion can be obtained with the use of an appropriate primer.

One-pack elastomers consist of polysiloxanes, fillers, catalyst, and cross-linking agent mixed together and packed into a sealed tube or cartridge. No curing takes place in the tube because the cross-linking agent is blocked by other groups, but when the material is applied and exposed to atmospheric moisture, the unstable blocking groups become hydrolysed off to leave reactive silanol (-SiOH) groups; a common example is acetoxysilane, which is hydrolysed with the release of acetic acid. The curing subsequently takes place from the surface inwards, with the rate partially depending on the rate of diffusion of moisture through the material. One-pack elastomers may cure incompletely in deep sections, but are very convenient for sealants and coatings. They are available in a range of viscosities from "non-sag" to "self-levelling", and they generally have good adhesion to non-primed clean substrates.

Condensation-cured RTV silicones find application in the main areas described below.

(i)   Encapsulation and potting. One of the biggest uses is the encapsulation or potting of electrical equipment in order to protect it from vibration and

potentially corrosive or contaminating atmospheres. Encapsulation imparts a
thin coherent overall coating to an item, whereas potting consists of embedding
components completely in the elastomer; by selecting a free-flowing RTV silicone, all spaces and crevices within an assembly can be filled. If necessary,
small entrapped air bubbles can be removed by placing the potted assembly under
a vacuum for a short time before the elastomer has cured.

Fig. 3.10. A two-pack silicone elastomer being used for potting a sensitive
electronic circuit.
Photograph courtesy: ICI Organics Division, Manchester, U.K.

Silicones are ideal for these electrical applications because of their excellent dielectric properties, which are retained over a very wide temperature
range, as are their other physical properties. It is also advantageous for
heat-sensitive electronic circuits that the curing takes place without exotherm,
and glass-clear elastomers are available if it is desirable that components
should still be visible. Faulty components can easily be changed by cutting out
the appropriate section of rubber, which is subsequently replaced by freshly-
catalysed rubber after the fault has been corrected. Typical applications for
encapsulation and potting are cable junctions, printed circuit boards, trans-
formers, dielectric elements, motor windings, etc. An encapsulating coating can
be applied by dipping, spreading, or spraying. For the last-named method, the
elastomer components are dispersed in an organic solvent, which is fed to a
spray gun; curing begins after the solvent has evaporated.

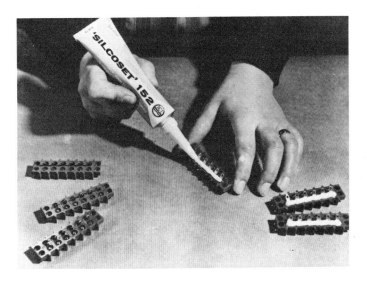

Fig. 3.11. A one-pack silicone being applied to terminal strips to form a moist-
ure-proof seal.
Photograph courtesy: ICI Organics Division, Manchester, U.K.

(ii) Sealing and caulking. Sealants represent an important application for
RTV silicones, particularly the one-pack type, due to their inertness (especially resistance to oxidation, corrosive chemicals, and ultraviolet light) plus
their retention of flexibility and adhesion at high and low temperatures. As
such, they are used in flanges, ducting, and in situ gaskets where service conditions preclude the use of conventional sealants, and various uses are found in
the aerospace, automotive and construction fields. These include seals for hot
air ducts, refrigerators and freezers, heat exchangers, and numerous parts of
industrial machinery; they have found widespread use as glass-to-metal seals for
windscreens, where the resilience of the seal is important for cushioning shocks
and vibrations. One-pack elastomers are employed in the construction industry
as seals for glazing and for joints between stone, brick, and concrete; two-pack
systems have found use in the U.S.A. as roof sealants, where the application of
the elastomer by brushing or rolling onto a primed roof surface results in a
continuous impervious layer. The long service life of silicone sealants often
makes them competitive in economic terms to other less costly alternatives.

(iii) Anti-stick coatings. Paradoxically, while RTV silicones can be used as
good adhesive sealants under one set of conditions, they can also be used as
anti-stick coatings under other conditions due to their excellent release characteristics with a wide range of materials. These include plaster, concrete,
polyester, epoxy resins, polyurethanes, waxes, and metal alloys, as well as

Fig. 3.12. Weather-proof silicone coatings are particularly suitable for protecting roofs of unusual configuration, as on this church.
Photograph courtesy: General Electric Company, New York, U.S.A.

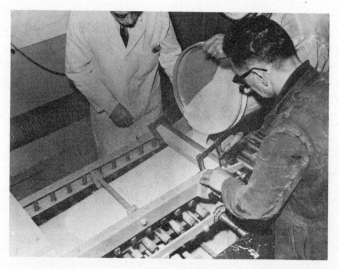

Fig. 3.13. A two-pack silicone being used to coat a belt.
Photograph courtesy: Dow Corning Ltd., Reading, U.K.

sticky foodstuffs such as pastries and chocolate. Anti-stick, heat-resistant coatings are commonly applied to conveyor belts, chutes and rollers that have to handle hot or sticky substances, and two-pack elastomers may be applied by knife coating or dip coating, the latter using the silicone rubber dispersed in a solvent. Fabrics such as cotton, nylon, and glass cloth are given this type of coating, and no priming is required since the elastomer penetrates into the fabric matrices to provide adhesion.

Paper coated with a silicone release agent finds numerous applications, such as backings for adhesive labels, tapes, and floor tiles, and linings for containers of wax and similar materials. Specially-formulated release papers are used for baking or packaging foodstuffs such as meats, pastries, sweets, frozen foods, meringues and cakes. Although the organotin catalyst dibutyltin dilaurate is classified as toxic by ingestion or skin absorption, it is non-migratory when incorporated into silicone coatings according to recommended procedures, and various proprietary release coatings are approved for use in food contact applications by statutory bodies such as the U.S.A. Food and Drug Administration.

In the paper-coating process, the silicone resin and catalyst are dissolved in a suitable solvent (e.g. hexane, toluene, etc.) and applied to the paper by techniques such as dip coating, reverse-roll, or gravure-roll coating. The

Fig. 3.14. Paper incorporating a silicone release coating can safely handle sticky foodstuffs.
Photograph courtesy: Dow Corning Europe Inc., Brussels, Belgium.

paper is then passed through an oven at about 120°C for up to one minute; this evaporates the solvent and enacts a fast cure of the silicone, which is usually stored for another 24 hours to develop the full release properties.

(iv) <u>Moulding and casting</u>. The release properties of silicone elastomers have led to their extensive use for producing moulds suitable for casting plastics, plaster, or low melting point alloys. The cured rubber is able to reproduce minute detail accurately, and a high tear strength, low shrinkage, and chemical inertness make possible hundreds of releases from a mould without any loss of definition in the moulded part. In a typical procedure for making a mould, the surface of a given master pattern or object to be reproduced is first thoroughly cleaned and then painted with a freshly catalysed silicone polymer; this ensures freedom from surface bubbles and a maximum reproduction of detail. The master pattern or object is placed in a mould box, and a catalysed two-pack silicone liquid is slowly poured in so that it completely covers the master. After the rubber has cured, the silicone mould may be easily separated from the master; because of its flexibility, it can be detached even when intricate undercuts exist in the pattern.

Silicone rubber moulds have become established in the production of texturised or embossed plastics, for example the leather-look PVC fabrics used for footwear. They are also used for the reproduction of a wide range of articles from engineering components to works of art and historical treasures. An exact impression of a person's jaw interior is obtained using an RTV elastomer so that precisely-fitting dentures can be made. An increasingly important use is for the moulding of furniture and other structural parts using integral-skin polyurethane foams.

(v) <u>Water-repellent treatments</u>. The hydrophobic properties of silicone polymers have been utilised for many years for the waterproofing of porous materials, principally masonry and textiles. These treatments, while making the materials water-repellent, still allow them to "breathe" by permitting the passage of air or vapours since the silicone does not seal up the pores of the material.

In the case of masonry treatments, two different systems based on silicones have been developed. One type utilises a water-based non-catalysed alkali metal siliconate solution; another type comprises a solvent-based silicone resin containing an organotin catalyst, which vulcanises on contact with moisture in a similar way to one-pack elastomers. The latter treatment uses a non-alkaline solution, and this is claimed to make it non-staining when used on natural stones containing iron. The solvent used for the resin may be xylene, white spirit, etc., and the solution may be applied by either brushing or spraying at close range; in either case, the water-repellent zone is formed at the surface, and does not penetrate right through the core of the stone or brickwork. However, if desired, a damp-proof course can be prepared by drilling a series of holes in a wall and then pressure-injecting the solution into them. A properly inserted silicone damp-course is claimed to have an effective life of much greater than 20 years.

Waterproofing is also commonly carried out on various types of textiles that are required to be weather-resistant, such as clothing, camping equipment, etc. Silicone bath treatments come in various forms, including aqueous- and solvent-based, and are designed for most common textiles, both natural and man-made. Most commercial silicone textile finishes are based on blends of polymers that cure by condensation, and tin-based catalysts such as dibutyltin dilaurate and stannous octoate are employed in many formulations, sometimes mixed with other less powerful catalysts such as zinc octoate or zinc naphthenate.

Having prepared the treatment solution according to the supplier's directions, it is applied to the textile by padding or by roller; for most fabrics, a silicone pick-up of 1-2% by weight gives the required degree of waterproofing. The fabric is then dried and the curing is completed by raising it to $130°-150°C$ for about 5 minutes. This results in a colourless cross-linked silicone film around each fibre of the fabric, which would produce a brittle feel to the material if only a single polymethylsiloxane-type polymer were used. However, by incorporating another "plasticising" silicone in the treatment, a softer finish is obtained that reproduces the original handle of the material. These condensation-cured systems can be applied from either aqueous emulsions, or solvents such as perchloroethylene, white spirit, or methyl ethyl ketone. The finishes are often designated as "permanent" by the manufacturer, and they show good durability to wear, washing, and dry-cleaning, although there is inevitably a gradual deterioration in water-repellency with long-term wear and repeated cleanings.

(vi)   Other special applications. As with any polymer system, the properties of RTV elastomers can be modified by incorporating other constituents. Thus while they normally have excellent thermal and electrical insulation properties, high conductivity silicone rubbers are manufactured by adding silver powder to the formulation; these may typically have a volume resistivity of about 10^{-3} ohm-cm, whereas the figure for a normal silicone rubber would be greater than 10^{14} ohm-cm. Conductive silicone sheet is used as seals, gaskets, O-rings,

etc., in various electrical applications where radio-frequency shielding is required, to insure either the containment or the exclusion of radio-frequency
energy; one example is as seals in waveguides to prevent the leakage of micro-
waves. As a combined hermetic and RF seal, the material retains its physical
and mechanical properties over the very wide temperature range characteristic of
silicone rubbers, typically -60° to +200°C.

Other examples of modified elastomers commercially available include products
containing glass fibre and P.T.F.E.

Silicone polymer systems can also be foamed by the inclusion of a blowing
agent in the formulation, together with a special catalyst to promote rapid cure
and gas evolution. The resulting closed-cell foam has excellent shock absorbing
and vibration damping characteristics, and can be used for potting exceptionally
sensitive electronic assemblies.

In addition to products comprising a blend of silicones with other materials,
there has been an interesting development recently where the cross-linking properties of silicones have been introduced into other polymers by silylating the
polymer chain; cross-linking can subsequently take place between the chains to
form a material with quite different properties. Dow Corning Corporation has
introduced Sioplas [*] E, a cross-linkable polyethylene that is supplied as a two
component system. One component comprises a silylated ethylene copolymer, while
the other is a masterbatch containing an organotin cross-linking catalyst.
Stored separately, each component is inert, but after mixing and exposing to
moisture, cross-linking takes place.

The components are mixed and processed by conventional means, e.g. extrusion,
injection moulding, etc. The fabricated part is then treated with water (usually at an elevated temperature) to cross-link the material. The resulting cross-
linked polyethylene has improved long-term strength, and is a non-melting plastic; it is claimed to show complete resistance to stress-cracking and to have
improved high-temperature performance and UV resistance.

In a related area to cross-linking via silicones, organotin catalysts have
been found to be very effective for use in conjunction with the organic silicate
binders employed in the production of refractory articles; this is discussed in
the next section.

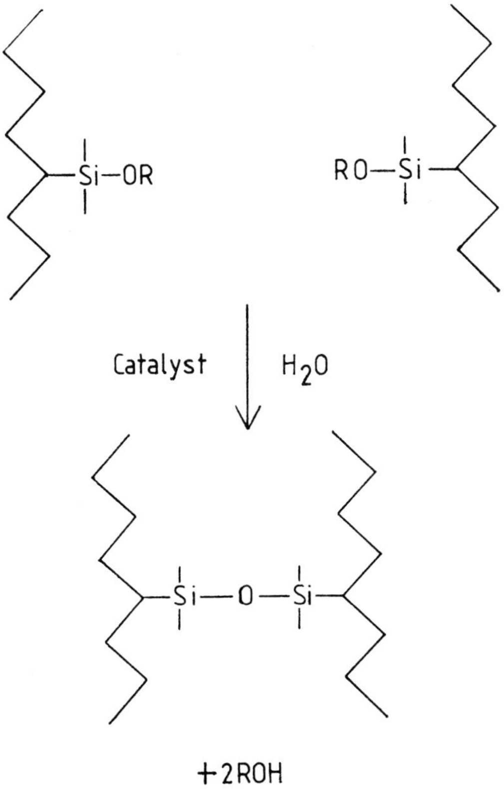

Fig. 3.15. Cross-linking polymer chains via siloxane formation.

3.3   ESTERS AND OTHER PRODUCTS

3.3.1   Industrial esterification processes

Esters constitute a very important class of industrial intermediates, and some examples have already been encountered in this and the preceding chapter in connection with polyurethanes and PVC. Some types of ester (including polymeric esters) and examples of their usage are shown in Table 3.4.

The esterification reaction can be represented by:

$$RCOOH + R'OH \rightleftharpoons RCOOR' + H_2O$$

Esters can also be produced by transesterification:

$$RCOOR' + R''OH \rightleftharpoons RCOOR'' + R'OH$$

TABLE 3.4

Esters and some of their applications

Ester	Applications
Polyester diols	Polyurethanes
Phthalates, trimellitates, adipates	PVC plasticisers
Unsaturated polyesters	Moulding resins, powder coatings
Alkyd polyester resins	Surface coatings, mouldings
Rosin esters	Food packaging coatings, cosmetics, food products
Fatty acid esters	Synthetic lubricants

Each of these reactions requires a catalyst and elevated temperature in order to obtain a good yield of the desired product. A range of tin compounds are marketed for use as esterification catalysts, including both organotins and compounds of tin(II). Inorganic and organic salts of tin(II) that are used in this field include the oxide, chloride, oxalate, glycoxide, and octoate, while organotin catalysts available include tetrabutyltin, butanestannonic acid, butyl-chlorotin dihydroxide, dibutyltin diacetate, and dibutyltin oxide.

When used as esterification catalysts, the tin(II) salts are insoluble in the ester product, and thus can be removed by filtration; the organotins, however, are soluble in the product and normally remain there unless it is felt particularly desirable to remove them. Tin-based catalysts are claimed to have several advantages over traditional acid catalysts (such as p-toluenesulphonic acid) in esterification and transesterification processes, namely:

a) A purer product is obtained since side-reactions are greatly reduced. Thus the need for post-treatments to remove aldehydes, ketones, and other coloured or odorous products is eliminated. There are also no ionic residues, which are common with acid-catalysed reactions.

b) Acid-induced corrosion of the process equipment (usually stainless steel) is virtually eliminated.

c) The tin-based catalysts are highly efficient, and need only be incorporated at levels of about 0.1 to 0.25%. The temperature required is of the order of 220°C, which is higher than that used in acid-catalysed processes (about 160°C), but the production cycle is faster.

An illustration of the high catalytic activity is shown in Table 3.5 which compares the degree of esterification attained in a standard reaction time (4 hours) for the reaction between 1.0 mole adipic acid and 2.2 mole 2-ethylhexanol; the esterification is indicated by the decrease in acid number of the reaction mixture (i.e. mg of KOH to neutralise 1g of sample).

TABLE 3.5

Comparison of catalysts for adipic acid/2-ethylhexanol; fusion method, nitrogen atmosphere, reaction time 4 hours [14].

Catalyst (0.15% w/w)	Temp. range	Final acid no.
None	200-220°C	23.7
p-Toluenesulphonic acid	128-170°C	18.9
Tetrabutyl titanate	180-220°C	2.8
Dibutyltin diacetate	170-220°C	0.4

The excess alcohol employed in these reactions is not degraded by the organotin compound, and can usually be recycled to subsequent batches without chemical treatment.

A survey of the principal organotin esterification catalysts produced and the applications for which they are particularly suitable is shown in Table 3.6.

TABLE 3.6

Organotin esterification catalysts and recommended applications

Catalyst	Suitable applications
Butanestannonic acid $[BuSn(OH)O]_n$	Esters such as phthalates, isophthalates, terephthalates, fatty acid esters; unsaturated and saturated polyesters, polyester diols, alkyd resins; transesterifications.
Butylchlorotin dihydroxide $BuSnCl(OH)_2$	Esters such as isophthalates, trimellitates, fatty acid esters; unsaturated and saturated polyesters, polyester diols; transesterifications.
Tetrabutyltin Bu_4Sn	Esters such as phthalates, adipates; esterifications requiring a mild catalyst.
Dibutyltin diacetate $Bu_2Sn(OOC.CH_3)_2$	Esters such as phthalates, adipates; alkyd resins; transesterifications.
Dibutyltin oxide $(Bu_2SnO)_n$	Esters such as isophthalates; saturated and unsaturated polyesters; alkyd resins; transesterifications.

3.3.2   Organic silicate binders

Organic silicates are widely used for the preparation of binders in the refractory and foundry industries, and in anti-corrosion paints based on zinc. The commonest example is ethyl orthosilicate, or tetraethoxysilane, $Si(OC_2H_5)_4$. This compound itself has no binding ability, but first it must be hydrolysed into a gel; this gel subsequently sets and acts as a binding agent for either the refractory or the paint. Ethyl silicate bonding allows the production of high quality refractories in alumina, sillimanite, mullite, zirconia, silicon carbide, etc., which have: a) reproducibility and high dimensional accuracy, b) uniformity of texture, and c) good resistance to thermal shock.

In practice, a technical grade of ethyl silicate is normally used that consists of a mixture of tetraethoxysilane and ethoxypolysiloxanes with up to five silicon atoms per molecule. It can be hydrolysed by either acid- or base-catalysed processes, but it has recently been found that a range of organotin compounds will catalyse its hydrolysis and gelation under neutral conditions [15, 16, 17]. It is stated that of the compounds examined, organotin alkoxides give especially transparent rigid gels; these compounds are not readily available commercially, but comparable results can be obtained by heating other organotins with ethyl silicate prior to hydrolysis. The addition of 5-20% of an organotin to ethyl silicate, with optional heating at 60°-160°C for up to $1\frac{1}{2}$ hours, causes the subsequent gelling to take place when water is added, and the conditions can be varied to give a carefully controlled gelling time in the range 20 min. to 10 hours. This control over gelling time is important in many applications.

Dibutyltin dilaurate and dibutyltin diacetate have been found to be particularly suitable for refractory and foundry applications because of the short gelling time induced, while bis(tributyltin) oxide is particularly useful in anti-corrosion zinc paints. In the latter application, not only is a high resistance to corrosion obtained, but the presence of bis(tributyltin) oxide endows anti-fungal properties on the paint. The zinc dust is added just prior to application, and the paint is touch dry in 10-30 minutes, and completely dry in 30-60 mins.

An example of the use with refractories is in the production of ceramic shell moulds by investment casting. In this process, an expendable mould of wax or other thermoplastic material is dipped into a slurry consisting of fine refractory plus ethyl silicate hydrolysate, and this coating is then dusted with coarse refractory. The coating is allowed to harden (about 20 mins.) and the sequence is repeated until a ceramic shell mould of the desired thickness is obtained; six coatings are adequate for a small mould. After leaving it to set overnight, the inner wax mould is removed and the ceramic article is fired.

The use of an organotin-catalysed ethyl silicate hydrolysate for bonding alumina refractories has been found to promote the added benefit of mullite formation, which is a desirable aluminosilicate phase that bonds the alumina particles together [15].

3.3.3   Olefin polymerisation and other reactions

Ethylene and other olefins can be polymerised under low pressure with catalysts that contain a tetraalkyltin or tetraaryltin compound in addition to aluminium trichloride, titanium tetrachloride or vanadium tetrachloride. Due to exchange reactions, the tetraorganotin acts as an in situ alkylating agent for aluminium trichloride, for example, producing a Ziegler-Natta polymerisation catalyst; the advantage of this method is that the organotin compounds are more stable and more easily handled than the corresponding aluminium alkyls.

While organotin compounds are not currently used on a large scale in this area, they have been suggested as catalysts or co-catalysts for this and several other processes, and the number of papers published and patents obtained with respect to organotin catalysts continues to grow steadily. A few selected examples are given in Table 3.7, but this is only a brief indication of possible potential applications.

TABLE 3.7

Examples of potential applications for organotin-based catalyst systems

Process	Organotin catalyst or co-catalyst	Reference
Ethylene polymerisation	Tetraphenyltin	[18]
Styrene polymerisation	Tetra-n-propyltin	[19]
Vinyl polymerisation	Tin mercaptides	[20]
Dye-sensitive photopolymerisation of vinyl monomers	Trialkylbenzylstannanes	[21]
Metathesis of unsaturated esters	Tetramethyltin	[22]
Copolymerisation of mono- and diisocyanates	Bis(tributyltin) oxide	[23]

REFERENCES

1 European Plastics News, August, 1979, p. 10.
2 Imperial Chemical Industries Ltd., Brit. Pat. 848,671.
3 General Tire and Rubber Co., Brit. Pat. 821,342.
4 E.F. Cox and F. Hostettler, Industrial and Engineering Chem., 52 (7) (1960) 609-610.
5 J.H. Saunders and K.C. Frisch, in Polyurethanes: Chemistry and Technology (I), Wiley-Interscience, New York, 1962, Chap. 4.
6 J.W. Britain and P.G. Gemeinhardt, J. App. Poly. Sci., 4 (11) (1960) 207-211.
7 S.L. Reegen and K.C. Frisch, in Advances in Urethane Science and Technology, II, 1973, Chap. 1.
8 T.E. Lipatova et al., J. Macromol. Sci. (Chem.), 4 (8) (1970) 1743.
9 Draka Square Block Process, Patent 96,048 (1958), The Netherlands.
10 Planiblock Process, Patent 424,434 (1973), Spain.
11 Maxfoam Process, Patent 131,636 (1970), Norway.
12 V.V. Servernyi, R.M. Minas'yan, T.A. Makarenko and N.M. Bizyukova, Vysokomol. Soedin., Ser. A, 18 (1976) 1276.
13 F.W. van der Weij, Makromol. Chem., 181 (1980) 2541-2548.
14 M and T Chemicals Inc., Technical Information Sheet 277, Catalyst T-1 (Dibutyltin diacetate).
15 H.G. Emblem and K. Jones, Trans. J. Brit. Ceram. Soc., 79 (4) (1980) 56.
16 Zirconal Processes Ltd., Brit. Pat. 1,494,209 (1977).
17 Zirconal Processes Ltd., Brit. Pat. 1,551,868 (1979).
18 W.L. Carrick, R.W. Kluiber, E.F. Bonner, L.H. Wartman, F.M. Rugg, and J.J. Smith, J. Amer. Chem. Soc., 82 (15) (1960) 3383-3387.
19 L.C. Anand, A.B. Deshpande and S.L. Kapur, Ind. J. Chem., 5 (5) (1967) 188-190.
20 Y. Nakamura, T. Ouchi, M. Imoto, Bull. Chem. Soc. Japan, 51 (12) (1978) 3574-3578.
21 D.F. Eaton, Photogr. Sci. and Eng., 23 (3) (1979) 150-154.
22 P.B. van Dam, M.C. Mittelmeijer and C. Boelhouwer, J. Chem. S. Chem. Comm., 22 (1972) 1221-1222.
23 S. Dati and A. Zilkha, Eur. Polym. J., 17 (1981) 35-40.

Chapter 4

WOOD PRESERVATION

4.1   INTRODUCTION

In spite of the many advances which have been made in materials technology, wood is still a valuable and much-used material of construction. Wood is easy and clean to work with and can be joined by gluing or nailing; it is suitable for varnishing or painting, has a relatively high strength/weight ratio and in thick sections is surprisingly fire-resistant. There is also the important consideration that wood is a renewable resource, originating from trees which can be planted and harvested. Yet this last factor also introduces a serious limitation, which is the susceptibility of all natural products to biodeterioration. Under certain conditions, wood can be attacked by fungi of various types and may also fall prey to wood-boring insects. The consequences may range from impairment of appearance to actual structural damage. Few buildings dating from before the 16th Century exist today. The cause of their demise has been largely fungal and insect attack, more destructive than the effects of fire, war or natural disaster.

Modern building methods tend to increase the risk of attack. Thinner structural sections are being specified in order to stretch existing resources and trees are being felled before reaching full maturity, thus increasing the proportion of susceptible sapwood at the expense of the more resistant heartwood. Improvements in living conditions have paradoxically introduced a fresh crop of problems. Central heating has resulted in higher temperatures in dwellings and this, together with the adoption of paint finishes which prevent the evaporation of moisture from damp external joinery timbers, often create the two conditions needed for decay to take place: dampness and warmth.

There are chemicals which will destroy these fungi and kill or repel insect pests, and preservative treatments based on these biocides have been devised. As will be described later, specialist companies are able to repair wooden structures which have been subject to decay, using suitable treatment solutions and whatever structural means are necessary. This need often arises when rot is discovered during routine inspections connected with buying or selling property. However, remedial treatment is often an expensive and time-consuming process and although the only solution for older or historic buildings, is best avoided in modern structures by pretreatment of the wooden parts before erection.

A suitable treatment has to meet several preconditions. The bioactive agent has to achieve sufficient penetration into the wood and must be uniformly distributed. It must be held firmly and losses by volatilisation must be minimal, not only from the point of view of continued protection, but also from the health and safety aspect. For similar reasons, the active agent must not be leached out by water, for example during heavy rainfall. European Standard EN 113 : 1980 covers the determination of the effectiveness of wood preservatives against wood-destroying Basidiomycetes, EN 73 : 1978 covers laboratory tests to assess volatilisation and EN 84 : 1979 deals with leaching tests. EN 20 : 1974, EN 21 : 1974 and EN 22 : 1975 describe tests for effectiveness of wood preservatives against various wood-boring insects. The wood preservative treatment must not materially alter the properties of the wood, i.e. its appearance, ability to be glued or painted, its flammability and (for precisely formed joinery) its dimensional properties.

Authorities in many countries now specify preservative treatments for wooden structures, for example the U.K. National House Builders' Registration Council specifies a standard immersion or impregnation cycle [1]. For pre-treatment of large quantities of structural timbers and wooden joinery, treatments which are organic-solvent based are employed, since they can be conveniently applied without leaving any harmful effects once the solvent has evaporated. The established ways of applying the preservatives are by brushing, spraying, deluging, immersion and impregnation.

4.2   ORGANOTIN COMPOUNDS IN WOOD PRESERVATION

Trialkyltin compounds were first proposed for use as wood preservatives in 1954 by van der Kerk and Luijten [2] who were carrying out a systematic investigation into the biocidal properties of organotins at the Institute of Organic Chemistry, TNO, Utrecht, Holland. As a result of these studies the high fungicidal activity of tributyltin compounds was uncovered.

4.2.1   Bis(tributyltin) oxide

In 1959 the compound bis(tributyltin) oxide was first introduced as a wood preservative by Osmose Wood Preserving Company in the U.S.A. The formulation contained the organotin in a base of colourless mineral oil and was applied by brushing, spraying or dipping. The product was marketed under the name "Oz", connecting not only with the company name, but also with their use of the well-known "Tin man" from the film "The Wizard of Oz" as trade name for the product (see Fig. 4.1).

Fig. 4.1. The original label referring to the first wood preservative to be based on an organotin compound.
Courtesy: Osmose Wood Preserving Company of America, Inc., Buffalo, NY.

Osmose is still marketing organotin-based wood preservatives today, which highlight the fact that there is now available 25 years of experience with this preservative. Bis(tributyltin) oxide soon became widely adopted as a wood preservative in many countries, its rather long chemical name being shortened commercially to "T.B.T.O.". ("TBTO" without full stops, is the trade name for the product as supplied by Albright and Wilson Ltd., one of the earliest manufacturers of the chemical.) T.B.T.O. is commonly employed at concentrations up to 3% in a solvent such as kerosene, usually in conjunction with a contact insecticide. Although still often applied by brushing or immersion, more sophisticated impregnation techniques have been developed for pre-treating joinery and other applications; these will be described in 4.6.

T.B.T.O. is a colourless liquid with a boiling point (at 2 mm mercury pressure) of 180°C, indicating low volatility; water solubility is also low, so that the compound is not readily leached from treated wood. It has a viscosity of 9 mPa.s at 25°C, a specific gravity of 1.17 at 20°C and a refractive index of 1.488 at 20°C. T.B.T.O. does not discolour wood which has been impregnated with it and the wood can be painted or glued in a normal fashion shortly after treatment; even light coloured paints can be used without any change in their colour. Flammability of the wood is not increased by the organotin once the solvent has evaporated. Although T.B.T.O. has a characteristic odour, this is not noticeable in treated wood at the concentrations normally used. The organotin has fairly low mammalian toxicity (acute oral toxicity, LD_{50}, for rats is about 200 mg/kg) [13] and follows the general pattern of tributyltin compounds in having a wide separation between the mammalian toxicity and fungicidal activity (see Chapter 10). It can be safely used provided reasonable safety precautions are adopted as described under 4.7. T.B.T.O. is the subject of a British Standard Specification B.S. 4630 : 1970, "Tributyltin oxide for use in paint, paper, timber and textiles as a preservative against microbiological attack".

Table 4.1 shows the biocidal activity of T.B.T.O. against various fungi and bacteria. As can be seen, it is extremely active against many fungi and gram positive bacteria although ineffective against gram negative organisms. The compound is compatible with many other biologically active agents and its biocidal activity is maintained even in alkaline solutions and in dilute mineral acids at room temperature.

TABLE 4.1

Biocidal properties of bis(tributyltin) oxide

Organism	Minimum concentration to inhibit growth completely ppm
Fungi	
Aspergillus niger	0.5
Chaetomium globosum	1.0
Penicillium expansum	1.0
Pullularia pullulans	0.5
Trichoderma viride	1.0
Candida albicans (yeast-like)	1.0
Gram-positive bacteria	
Bacillus mycoides	0.1
Micrococcus pyogenes var. aureus	1.0
Bacterium ammoniagenes	1.0
Gram-negative bacteria	
Pseudomonas aeruginosa	> 500
Aerobacter aerogenes	> 500

Although T.B.T.O. has some insecticidal activity, it is not a powerful contact insecticide and other agents may be used to reinforce this activity in commercial formulations. These include chlorinated hydrocarbons such as gamma benzene hexachloride and some synergistic action has been observed between the two biocides [4]. Synthetic pyrethroids such as "Permethrin", which have high toxicity against beetles and termites, have been used in combination with

T.B.T.O. as has dichlorofluanid [5]; the latter is reported to be effective against blue-rot and other surface fungi which spoil the appearance of wood and to have a synergistic effect with T.B.T.O.

Fig. 4.2. Examples of untreated timber (left) and timber treated with T.B.T.O. (right).

Water repellents are sometimes included in preservative formulations since they reduce the rate at which the wood absorbs water, causing it to run off, rather than soak in or be absorbed by the end grain. In this way they reinforce the effect of the preservative by keeping the wood dry and reducing dimensional changes.

4.2.2   Other tributyltin compounds

Tributyltin phosphate is employed in at least one commercial wood preservative and is said to offer the advantages of low mammalian toxicity, very low vapour pressure and a flash point greater than 55°C. Tributyltin benzoate is used in some countries in place of bis(tributyltin) oxide in order to meet legislative requirements. Tributyltin compounds formed by reaction between T.B.T.O. and a long-chain carboxylic acid also have some merit for wood preservation, for example tributyltin linoleate, naphthenate and abietate. Physical properties of various tributyltin compounds are shown in Table 4.2.

The long chain carboxylates have lower water solubility than T.B.T.O. (Table 4.3) and therefore greater leach resistance. They are also less volatile than

T.B.T.O. (Table 4.4).

Organotin polymers based on tributyltin methacrylate have also been tested as wood preservatives and are described in 4.8.2.

It should be borne in mind that the formulation of the preservative may influence its effectiveness and the choice of solvent, binder, siccative and colourant, etc., is important. Although the biocidal activity is unaffected as long as the tributyltin configuration remains intact, other properties such as storage stability, working properties and wood penetration may be affected. The activity of individual formulations is tested by the manufacturer using standard tests.

4.2.3 Laboratory testing of fungicides

A number of fungal tests have been devised for measuring the effectiveness of a preservative in the laboratory. In these tests the activity of a biocide is assessed in a particular kind of wood against one organism. Through a series of tests, a picture can be built up of biocidal activity against a group of organisms in a range of wood types. These tests are very empirical, since a large number of variables can influence the results. Different amounts of preservatives are necessary to protect different woods and when comparing two preservatives, the relative performances will vary according to the wood which is treated. For preliminary screening, a preservative may be incorporated in agar medium containing the fungal strain, and inhibition of mould growth after incubation assessed. However, the organic-solvent-borne preservatives have to be emulsified into the agar and the method of emulsification has been found to influence the results obtained, so that such techniques are not used for any detailed assessments [6].

The normal method consists of impregnating small standard wood blocks with the test biocide incorporated in a suitable organic solvent. A suitable laboratory test set-up can be used to achieve vacuum impregnation and wood samples are treated with the test solution at a graded series of concentrations of the biocide. The wood blocks are dried, weighed, sterilised and exposed to pure cultures of the test organism growing on malt agar. After a suitable incubation period (typically 6 weeks at 22°C), the blocks are removed, superficial fungal growth removed by gentle scraping and the blocks dried and re-weighed when cool. The degree of decay is assessed from the loss in weight. Sometimes soil is used as the nutrient medium. A strip of wood, inoculated with the test organism, is placed on the surface of the soil and when fungal growth has been well established, test blocks impregnated with the biocide are placed on the

TABLE 4.2

Physical properties of tributyltin preservatives [5]

Property	T.B.T.O.	TBT benzoate	TBT linoleate	TBT naphthenate	TBT abietate
State	liquid	20°C liquid	liquid	liquid	liquid
Colour	clear	clear	yellow	dark yellow	yellowish brown
Viscosity (mPa·s at 25°C)	ca. 9	ca. 20	ca. 500	ca. 500	ca. 700
Density (g/ml at 20°C)	ca. 1.17	ca. 1.19	ca. 1.04	ca. 1.09	ca. 1.11
Flash point (°C, Pensky-Martens)	190	110	105	78	89
Sn content (% by wt.)	ca. 39	ca. 28.5	ca. 21	ca. 22	ca. 19

TABLE 4.3

Tap water solubility of tributyltin compounds [5]

Compound	ppm tin
T.B.T.O.	30
TBT benzoate	2.4
TBT linoleate	1.0
TBT naphthenate	1.3
TBT abietate	1.2

TABLE 4.4

Volatility of tributyltin compounds [5]

Compound	Weight loss after 144 h at 65°C in dry air
T.B.T.O.	26
TBT benzoate	8.5
TBT linoleate	1.0
TBT naphthenate	6.0
TBT abietate	2.1

wooden strip and incubated as before. Details of the wood block/agar test are given in British Standard B.S. 6009 : 1982 (EN 113). The soil block technique is specified in the U.S. Standard A.S.T.M. D 1413. Results can be interpreted in terms of a "toxic limit", which is the interval between the concentration of toxicant and its associated loading in wood which permits significant decay and that which prevents decay. "Significant decay" is considered to be an average weight loss of at least 3% (or at least 5% for an individual block). Sometimes "threshold values" are calculated; these are obtained by extrapolating a graph of weight loss versus treatment level to give a single concentration at which no weight loss occurs. This has a slight limitation in that it implies a mathematical significance for the value, which it does not actually have. Results of all wood block tests must be carefully interpreted; they are most suitable for use with Basidiomycete fungi and are less suitable for soft rots.

Leaching tests are sometimes applied to treated wood samples prior to testing in order to assess whether any loss of effectiveness results. A leaching procedure is prescribed in British Standard B.S. 5761 : Part 2 : 1980. In general terms, this test consists of vacuum impregnating wood blocks, followed by immersion for 14 days with 9 changes of water, followed by drying for 14 days.

Using a modification of the wood block/agar test method with small test blocks (30 x 10 x 5 mm) to reduce the incubation period, a large number of tributyltin compounds have been screened for fungicidal action in the laboratories of the International Tin Research Institute. Results are summarised in Table 4.5. Since one aim of the tests was to compare the toxicity of particular tributyltin compounds with that of T.B.T.O., solution concentrations of preservative were adjusted to give tributyltin loadings in wood corresponding to those obtained with T.B.T.O. at, above and below, the published toxic level of T.B.T.O.

TABLE 4.5
Toxic limits of tributyltin compounds against two wood-destroying fungi [7]

Compound	Fungus	Toxic limits[*] (loading in wood blocks kg m^{-3})
T.B.T.O.	C.p.	0.19-0.70
	C.v.	0.20-0.69
Tributyltin borate	C.p.	0.38-0.96
	C.v.	0.19-0.37
Tributyltin carbonate	C.p.	0.20-0.38
	C.v.	0.20-0.38
T.B.T.O./glucose	C.p.	0.46-1.14
	C.v.	0.47-1.17
Tributyltin nitrate	C.p.	0.48-1.13
	C.v.	0.45-1.16
Tributyltin phosphate	C.p.	0.38-0.96
	C.v.	0.39-0.99
Tributyltin ethanesulfonate	C.p.	0.49-1.21
	C.v.	0.24-0.50
	C.p. ≠	0.58-1.18
	C.v. ≠	0.58-1.17
Tributyltin sulphide	C.p.	0.40-1.03
	C.v.	0.39-1.00
Tributyltin chloride	C.p.	0.38-0.97
	C.v.	0.39-0.98
	C.p. ǂ	0.16-0.31
	C.v. ǂ	0.31-0.61

C.p. = Coniophora puteana   C.v. = Coriolus versicolor

[*] Mean values from several tests

≠ Applied in water   ǂ Applied in acetone
In all other cases the solvent was petroleum ether.

A significant finding was that the biocidal activity seemed to be little influenced by the nature of the group attached to the tributyltin radical.

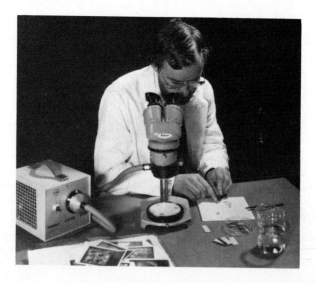

Fig. 4.3. Testing of organotins as wood preservatives. Sectioning impregnated wood samples at the International Tin Research Institute.

4.3   MECHANISMS OF FUNGAL ATTACK

The area where organotins play their major part in wood preservation, is in preventing attack by wood-destroying fungi. Before considering the types of fungi and the way in which organotins exert their fungicidal activity, it will be helpful first to look briefly at the structure of wood.

4.3.1   Structural aspects of wood

The source of wood is the tree, which itself is a plant form which has adopted a mode of growth involving production of "woody" tissues which enable it to reach a greater height towards the light than "non-woody" plants. The woody tissues carry out vital functions for the tree in transporting sap and storing food reserves. As the wood tissue ages, it loses the ability to convey sap, but forms a core of high mechanical strength. The trunk is the portion of the tree involved in the production of structural timbers. A central pith is surrounded by a series of concentric cylinders; wood tissue around the pith is termed heartwood and consists of cells which have died. Then come the living cells of the sapwood, or xylem which in turn is covered by a thin layer of phloem cells and protective bark. Between the xylem and the phloem cells is a very thin sheath, only a few cells thick, known as the cambium. This is the actively growing area of the tree; dividing cells are passed either inwards to join the xylem, or outwards to form the phloem. As the diameter of the trunk increases, the protective bark layer is pushed outwards and becomes stretched, cracked and ridged in a characteristic pattern.

In temperate climates, growth tends to be seasonal and varies from a wide band of large but thin-walled cells (early growth) to much smaller thick-walled cells (summer or autumn growth). The transition zone between these cell regions appears in cross section of the trunk as a discernable ring; the number of rings indicates the years of growth and hence the age of the tree. The xylem serves to conduct water and dissolved salts from the roots to the outermost parts of the tree. Phloem cells conduct sugars from the crown to the various regions of the tree.

When a xylem cell is first formed, it consists of a thin wall made up of sugars which have polymerised into cellulose. Sugars from the phloem continue to be supplied to the xylem cells so that successive secondary layers of cellulose are formed within the original primary wall. Once the cell structure is complete, much slower processes of lignification occur, consisting of a progressive deposition of lignin initially within the middle lamella (an amorphous, undifferentiated region between cells) and then within the cell walls. This serves to stiffen and strengthen the cells. A schematic interpretation of the cell structure is shown in Fig. 4.4.

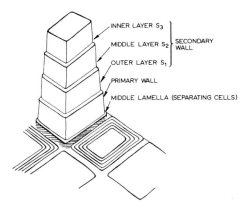

Fig. 4.4. Structure of a wood cell. Cells are 1-5 mm in length and about 95% are aligned along axis of tree trunk. Cell walls are made up of: 45-50% microfibrils of a crystalline cellulose polymer containing 5000 glucose units; 20-25% semi-crystalline non-cellulosic polysaccharides and lignin amorphous 3-dimensional molecules of phenolpropane units.

A comprehensive treatment of this subject and of wood preservation in general has been presented by Richardson [8]. The inactive cells formed by changes in the living cells of the inner sapwood make up the heartwood. This contains a large amount of waste matter (arising from living processes within the tree) which reduces the porosity and is often toxic thus making the heartwood more resistant to insect and fungal attack and to swelling and shrinkage with changes in moisture content. Relative quantities of heartwood and sapwood are regulated by dynamic living processes within the tree. The practice of cutting down trees before they have reached full maturity means that there is a greater proportion of less resistant sapwood in the timber so produced.

4.3.2 Fungal attack

For fungal attack of wood to occur, the wood must have a high moisture content and have access to air. Fungi produce an enzyme, cellulase, which breaks down cellulose and brings about decomposition of the cell wall components. According to some theories [9, 10] bacteria are initially responsible for the breakdown of pits in the sapwood, opening up the wood structure and making it more permeable to liquids and gases so that fungal growth is possible. Primary moulds then enter sapwood cells through these initial pathways. Primary breakdown products are partially absorbed by the fungus and serve as a source of nourishment for it.

Fungi consist of a number of very slender branched, thread-like cells, known as hyphae. The mass of interlacing hyphae is known as a mycelium and this is the vegetative part of the body. When the fungal mycelium has accumulated sufficient food reserves, a fruiting body, or sporophore, is produced, consisting of hyphae tightly packed together and with a characteristic shape and colour. Spores are minute structures invisible to the eye but in large quantities appearing as fine dust. The dry rot spore measures only about 0.001 mm and weighs 1×10^{-11} g; this low weight enables the spore to remain suspended in an air current for a very long time and therefore is easily distributed. It has been estimated that a sporophore of the dry rot fungus one yard square could produce 50 million spores per minute over many days [11].

The first stage of fungal attack consists of invasion of the cellular passages in the wood by hyphae, originating from spores, which progress via natural pits in the wood from initial pathways into adjacent tissues. There is an initial period of passive vegetative growth when the wood substrate is rapidly and extensively colonised. The hyphae then grow preferentially through the largest available voids in the substrate and into the nutrient-rich tissues containing simple sugars. In the initial stages hyphae grow from cell to cell, through

natural pit openings rather than by boring through lignified cell walls. Extra-
cellular secretions actively degrade the components of the cell wall and these
lysis patterns are determined by the type of invading fungus and by the nature
of the secretions. Physical and chemical differences in the cell wall structure
of particular types of wood also modify these lysis patterns. These phenomena
are manifested to varying extents in the types of behaviour observed in the different processes of wood rot.

4.3.3   Principal types of fungi

Major types of wood-destroying fungi are summarised in Table 4.6; there is an
enormous range of fungi which will attack wood, but dry rots and wet rots are by
far the most significant, accounting for some 95% of all fungal decay in the
U.K. The wood-rotting fungi can conveniently be divided broadly into brown
rots, white rots and soft rot fungi; there are also fungi which, whilst not
causing any structural damage, do affect the appearance of the wood, for example
the blue stain fungi. Brown rots attack only the cellulose in the wood and the
predominant enzymic action is hydrolysis of polysaccharides. Typical examples
are the well-known dry rots and wet rots.

A number of fungi can cause dry rot, but by far the most important is
Merulius lacrymans. The term "dry rot" is descriptive of the dry and friable
condition of the rotted wood, since like other fungi it requires moisture to
flourish. One of its most troublesome characteristics is the ability to produce
water-carrying strands or rhizomorphs which develop from hyphae and are able to
pass over and even through masonry, seeking fresh wood to infect; this makes
eradication difficult. Any dry wood encountered by the hyphae is first moisten-
ed and then attacked. The decayed wood is brown in colour and has the appearance of being charred, with cuboidal cracking. The specific name "lacrymans"
(weeping) refers to the fact that innumerable globules of water may be seen when
the fungus is in active growth, which sparkle like teardrops when illuminated.
When the fungus is exposed to light, fruiting bodies appear, which are thin and
pancake-like, white around the edges (Fig. 4.5). Spores are rusty-red in colour.

Wet rot, Coniophora puteana or cellar fungus, is the next most important
wood-destroying fungus in buildings. It depends for its existence on continual
dampness or running water to keep the wood damp. Although more prevalent than
dry rot, its effects are not usually so severe. Fungal strands are not as thick
as those of Merulius lacrymans and are vein-like in appearance when growing on
the surface of wood or plaster. Sporophores are rarely found indoors but the
spores are widely present and wood of the required moisture content runs a high

TABLE 4.6

Common fungal causes of timber decay

Fungus	Common name	Characteristics
Merulius lacrymans	Dry rot	This fungus requires moist, warm, still conditions for growth. Dry rot spores will germinate when accidental wetting of the wood, followed by partial drying, permits the relative humidity to reach an optimum level. Hyphae (rootlings) penetrate into the wood and spread, resulting in a soft white growth, like cotton wool. Affected wood has a brown colour and a dry, powdery appearance, from which the rot gains its name. Cuboidal cracking of the wood is another characteristic. The rot can penetrate masonry which makes complete eradication difficult.
Coniophora puteana	Wet rot (Cellar fungus)	Requires more moisture to thrive than does dry rot fungus and depends on continual dampness or running water to keep the wood damp. It often occurs in cellars or in roofing timbers. The early growth has a fluffy, white appearance, often developing into a stringy shape with fine strands brown or black in colour. The wood may also darken in colour and crack lengthwise.
Poria vaporaria	Wet rot (Pore fungus)	Feeds on damp softwood, causing its ultimate collapse. It tolerates more extreme temperature conditions than the dry rot fungus. Its hyphae are usually formed in a fan shape over the timber surface and are snow white. The fungus is common in damp buildings.
Coriolus versicolor	White rot	Occurs in hardwoods, especially when there is ground contact, as in pit props. The sporophore is rarely seen, but consists of a thin bracket up to 75 mm across, grey and brown on top, with concentric hairy zones and a cream pore surface underneath from which spores are released. Decayed wood is lighter in colour and much weaker than unharmed wood.

Fig. 4.5. Fruiting body of the dry rot fungus <u>Merulius lacrymans</u>.
Photograph courtesy: Wykamol Ltd., Winchester, U.K.

risk of being attacked in buildings. White rot fungi attack both lignin and
cellulose in wood, oxidising enzymes being responsible for attacking the lignin.
Decayed wood differs from that attacked by dry or wet rot, being darker in colour and with longitudinal cracking. <u>Coriolus versicolor</u> is the commonest cause
of white rot in hardwoods, especially in ground contact such as occurs with pit
props. The sporophore is not often seen; decayed wood appears to be lighter in
weight and much weaker than unattacked wood.

Soft rot fungi may attack wood which is waterlogged and are particularly prevalent on the wooden fillings of water-cooling towers. The fungi are primarily
cellulose destroyers and attack is usually superficial, so that except in thin
sections there is little weakening action. Softwoods as a class are more resistant to soft rots than are hardwoods, possibly because of the higher degree of
lignification of the secondary cell wall in softwoods which protects the cellulose from attack. Recent work [9] on the mechanism by which fungi decay wood
has been concerned with attempts to understand how the decay patterns of brown,
white and soft rot fungi occur in terms of their respective enzyme activity.
Evidence has been produced to demonstrate that the enzymes of the soft and white
rot fungi are restricted in their activity to regions of close proximity to the
hyphae, whereas those of the brown rot fungi appear to be acting at a much
greater distance. It has therefore been postulated that the enzymes of soft
and white rots are linked to the hyphae, whereas those of the brown rots are

not.

Sapstain fungi, so-called because the staining is usually confined to sapwood, can cause problems not only with felled timber which may depreciate in
quality due to unsightly appearance, but also with disfigurement of decorative
wood cladding on buildings. The wood surface may be actively colonised by fungi
causing a rapid greying (blue stain) which require only occasional wetting for
growth; Northern hemisphere softwoods are particularly susceptible [12]. The
predominant species have been shown to be Pullularia pullulans and Cledosporium
herbarum. The growth of P. pullulans on exposed timber is characterised by the
presence of slimy exudates on the surface of the hyphae which probably assist
adhesion and prevent dessication during drought. The hyphae are brown when examined under the microscope, the blue colour resulting from refraction of light
[13]. Sapstain is almost invariably accompanied by superficial discolouration
caused by moulds including common genera such as Penicillium, Aspergillus and
Trichoderma.

4.4   FUNGICIDAL MECHANISMS OF ORGANOTINS

The anti-fungal activity of bis(tributyltin) oxide as determined in vitro
differs from the actual behaviour exhibited in wood samples against the same
fungal species. This would be expected to some extent, since factors such as
the type of wood treated, depth of penetration and uniformity of distribution of
preservative, other constituents of the formulation and the long-term stability
of the toxicant in the wood, all play a part. Although organotin compounds have
given long-term protection to many types of wood in various situations and environments, there are certain aspects of the mechanism by which this protection
is exerted which are not clearly understood. Tributyltin compounds are most effective against brown rot fungi, which destroy cellulose, but rather less effective against the white rots, which attack both cellulose and lignin. The preservative action seems to be associated with cellulose rather than with other
wood components. A combination of T.B.T.O. and pentachlorophenol has been shown
to give excellent preservation against white rots, the tributyltin evidently
protecting the cellulose and the pentachlorophenol protecting the lignin [14].
Pentachlorophenol alone has little action against white rot, probably because
white rots can attack polyphenols such as lignin and are therefore equally tolerant towards phenolic preservatives. Energy required for phenol degradation by
the white rot is probably supplied by the attack on cellulose and when the cellulose is protected by tributyltin, this energy is not available and the phenolic
biocide can exert its protective effect.

One interesting factor is the apparent independence of the fungicidal activity of a large series of tributyltin compounds Bu_3SnX on the nature of the group X. Various possible explanations have been attempted [7]. One possibility is that the tributyltin compounds are hydrolysed to T.B.T.O. in the wood; however, most of the compounds studied had hydrolytically stable Sn-X bonds and hydrolysis in the presence of a wood substrate is unlikely. ^{119m}Sn Mössbauer studies at the International Tin Research Institute [15], using Scots pine samples, demonstrated that bis(tributyltin) carbonate is formed very rapidly when T.B.T.O. is impregnated into Scots pine, due to an in situ reaction with carbon dioxide trapped in the wood. This compound has very similar toxic limits in wood to those of T.B.T.O. There is a possibility that the tributyltin compounds all form the carbonate but no reactivity towards carbon dioxide was shown by the tributyltin compounds when they were dissolved in toluene and the gas bubbled through the solution.

Richardson and Lindner [16] have suggested that the preservative acts by protecting active sites on wood cellulose, rather than by a direct biocidal action on the fungus. This would imply that the organotin is bound to these active sites. The authors postulated that the organotin condensed on the hydroxyl groups at the terminal 1 and 4 positions. If tributyltin carbonate is formed, hydrogen bonding between this carbonate and the wood cellulose remains a possibility but this would not be expected to involve only the terminal 1 and 4 hydroxyl groups.

Another possible explanation for the anion independence of the preservative activity of tributyltin compounds, is related to the fact that the toxic action of tributyltins against fungi appears to be intra-cellular. These compounds have been shown in mammals to be strong inhibitors of oxidative phosphorylation in mitochondria [17] and it has been proposed that the same mechanism is operative in fungi and that thiol (HS-) groups are involved. In the simple Bu_3SnX compounds, exchange of the X groups at active protein sites may proceed with relative ease and hence no significant variation in fungi toxicity with X would be observed.

Detailed electron microscope studies in beechwood by Bravery [18] also suggest that the primary mode of toxic action of tributyltins against fungi is intra-cellular in nature. The preservative was seen to be distributed uniformly in the wood vessels, but more irregularly in the adjacent fibres. Coriolus versicolor was able to grow into the vessels and fibres although the initial rate of penetration was retarded and the fungus was able to produce extra-cellular secretions despite the presence of preservative. Erosion of the preservative deposit was observed as well as lysis of underlying cell walls and denaturing of the hyphae. The indications were that solubilisation of the preservative would probably be necessary before the toxic components could be released and then absorbed by the hyphae to kill them. It has been observed [19] that in test blocks with non-uniform distribution of T.B.T.O., the fungus Coniophora puteana can pass through areas of high preservative loading to those of lower loading where decay can occur. This would suggest that the preservative is not readily available to the hyphae in the loaded areas. It has been suggested [9] that a prerequisite for fungal attack may be a colonisation of hardwoods and softwoods by groups of micro-organisms and the presence of a wood preservative serves to delay this colonisation.

An interesting body of work has been concerned with the use of preservatives in solvents which induce swelling in the wood and the resultant effect on fungicidal activity. Bravery [20] has shown that when T.B.T.O. is incorporated in an organic solvent which induces swelling (such as dioxan + 5% water), the preservative is deposited not only in the cell interior but also within the thickness of the S_2 cell wall layer (see Fig. 4.4). Greater uniformity of distribution between the different wood elements is also achieved. Toxicity testing indicated that the white rot Coriolus versicolor was inhibited by lower levels of T.B.T.O. than when a non-swelling solvent was used. The improvement in toxic action could thus be due to the greater resistance of T.B.T.O. deposits to enzymatic removal when they are "keyed" into the S_3/S_2 wall region than when they are only deposited superficially. In addition, if active protein sites in the fungus do in fact involve free thiol groups, as discussed earlier, then T.B.T.O. may combine more readily with these in an aqueous medium than in an organic solvent and this has been observed for compounds containing thiol functions.

Cox [21] studied a number of solvent systems to see whether enhanced fungal activity could be obtained for T.B.T.O. in wood by using a solvent which would cause the wood to swell slightly but not enough to render it commercially unacceptable on account of dimensional change. Only the dioxan + 5% water system described earlier, gave improved activity; swelling solvents such as methanol did not influence the fungicidal properties. Varying the moisture content of the wood was found to have no effect on preservative action. However, the author considered that with some wood-destroying fungi, such as the soft rots, effective action might be obtained if there were significant penetration of the cell wall by T.B.T.O. with solvents containing water-miscible components such as 2-ethoxyethanol.

Attempts have been made to develop preservative treatments which are effective against sapstains. Coggins [22] has tested a large number of compounds, including combinations of organotins with quaternary ammonium compounds, against sapstain occurring in wood during seasoning or in storage after conversion. The aim was to find an alternative to the effective but environmentally unsatisfactory sodium pentachlorophenate. Best results were obtained with a mixture of quaternaries without an organotin compound and in fact organotin compounds have not proved particularly successful in this area. It is reported, however [5] that tributyltin compounds used in combination with dichlorofluanide exert a synergistic biocidal action on the blue stain fungi Pullularia pullulans and Sclerophoma pityophila.

4.4.1   Stability of organotins in wood

In 1978 workers at the Princes Risborough Laboratories of the Building Research Establishment in the U.K. reported evidence of degradation of tributyltin to dibutyltin in situ in treated wood. This would imply a decrease in protective action since dibutyltin compounds are generally less biologically active than the corresponding triorganotin. Since then, a number of workers have reported degradation of tributyltin compounds in experimental studies. Jermer et al. [23] examined window frames, which had been impregnated with T.B.T.O., after five years of service, for evidence of fungal attack and degradation of the tributyltin. High performance thin layer chromatography was used to separate degradation products of the organotin; a substantial degradation of tributyltin to dibutyltin and monobutyltin was evidenced. This was greater nearer the surface, suggesting that temperature may be important for degradation. The bioassay indicated that resistance to decay decreases with the breakdown, 0.05-0.06% T.B.T.O. remaining in wood seeming to be the minimum level for effective protection. It was considered that the presence of several fungi capable of causing detoxification might have contributed to the breakdown. These workers also looked at the effect of accelerated solvent evaporation by heating treated wood in kilns at 30-40°C on the distribution and degradation of tributyltin naphthenate in vacuum impregnated wood [24]. Forced evaporation caused a loss of organotin from the surface layer, leaving a fairly uniform distribution of preservative as a function of distance from the surface, whereas in naturally evaporated samples there is a fairly steep gradient in concentration. Substantial degradation of the tributyltin naphthenate was indicated, but samples heated in the kiln did not show any greater degradation than those stored at room temperature and allowed to dry. Degradation was thought to result possibly from reaction with the wood after impregnation.

Plum and Landsiedel [25] studied the effectiveness of carboxylic acids, higher alcohols and aldehydes as stabilisers of T.B.T.O. against the effects of UV light. Table 4.7 shows inhibition of growth of Coniophora puteana obtained with unstabilised and stabilised T.B.T.O. after 48 hours exposure to UV. In further studies T.B.T.O. containing 1 or 3% of a stabiliser was impregnated into spruce wood logs which were half buried in the soil for 9 years, exposed to sunlight and the atmosphere.

TABLE 4.7

Biocidal activity of T.B.T.O. after UV irradiation [25]

	Inhibition zone of Coniophora puteana	
	Before irradiation mm	After 48 h UV mm
T.B.T.O. unstabilised	12	5-8
T.B.T.O. stabilised	10-13	10-12

Table 4.8 shows the relatively small amount of degradation of T.B.T.O. after exposure. In biocidal studies with various fungi, samples with the higher stabiliser level showed best results.

TABLE 4.8

Degradation of T.B.T.O. in soil/sunlight tests [25]

		Degradation product after 9 years wt. %		
Compound		Monobutyltin	Dibutyltin	Tributyltin
T.B.T.O. + 1% stabiliser	Airzone	18	32	50
	Earthzone	8	26	66
T.B.T.O. + 3% stabiliser	Airzone	6	17	77
	Earthzone	7	15	78

It is interesting to note that if the organotin compound confers protection by binding to reactive sites on cellulose then the protective action would not be adversely influenced by some degradation. It has been reported [26] that organotin preservative in treated wood was still conferring decay resistance after 9 years exposure in Hawaii, in spite of the fact that 85-90% of the tributyltin oxide had been dealkylated to dibutyltin, monobutyltin and inorganic tin.

The mechanisms of fungal attack are fully covered in a review up to 1981 by Evans and Hill [27].

4.5 PROTECTION AGAINST INSECT ATTACK

4.5.1 <u>Wood-boring insects</u>

In addition to fungal attack, wood is also susceptible to damage by wood-boring insects which can burrow into it, extracting the very small proportion of protein present and either digesting the cellulose or utilising starchy cell contents as an energy source. Prior fungal attack will convert the wood into a form readily assimilated by the insect and in some cases this is a prerequisite before the wood-borer can enter the wood. Attack by insects is insidious in that often the only external signs of infestation are a few small holes, whereas internally the wood structure may be extensively damaged. The general pattern of insect attack is that eggs are laid in small cracks or blemishes in the wood surface and these hatch out as larvae which immediately begin boring into the adjacent wood. During its larval stage, which may be several years in duration, the insect feeds on the wood; the full-grown larva then forms a chamber just below the surface in which it pupates into an adult beetle which bores its way to the surface and escapes via a flight hole.

Typical wood-destroying insect borers are illustrated in Fig. 4.6. One of the most widespread timber-attacking pests, particularly in Britain, is the

THESE INSECTS SPELL...

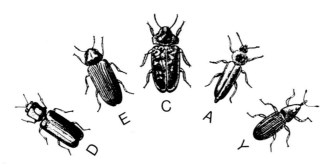

Fig. 4.6. Wood boring insects, left to right (with actual sizes) : Powder Post Beetle (5 mm), Common Furniture Beetle (4 mm), Death Watch Beetle (8 mm), House Longhorn Beetle (19 mm), and wood weevils (5 mm).

Common Furniture Beetle, <u>Anobium punctatum</u>. This will attack most well-seasoned timbers and its life cycle is about 3 years. Eggs are laid in cracks and crevices, larvae hatch out by breaking through the base of the egg and boring immediately into adjacent wood. The insect remains in the larval stage for about a year and when full grown the larva is a white curved grub 0.6 mm long with a brown head and strong biting jaws. It bores to within a small distance from the surface of the wood and pupates. After a few weeks in the pupal stage the adult beetle bores to the surface and escapes via a circular flight hole about 1.5 mm in diameter. The beetle then mates and lays eggs once more, repeating the cycle. The beetle is dark brown in colour and has lines of small punctures on its wing covers from which it derives its proper name. The bore dust is coarse with a sandy feel.

Damage by the Death Watch Beetle, Xestobium rufovillosum is most frequent in older buildings where oak roofing and other structural timbers may be attacked; softwoods are attacked only rarely. A prerequisite for attack is the prior occurrence of fungal growth. Larvae hatch after 2-8 weeks and explore the wood before starting to bore down into it. Larvae may remain in the wood for anything from a year and a half up to several years and they emerge leaving a flight hole about 3 mm in diameter. Larvae are larger than those of the furniture beetle and are covered with long yellowish hairs. The bore dust is characteristic in that it contains large bun-shaped pellets. The beetle gains its name from the ticking sound made by the beetle rapping on wood with its head during the mating period.

The third species of wood boring insect of major significance in buildings is the House Longhorn Beetle, Hylotrupes bajulus. This is the largest of the woodborers and the larvae feed voraciously on the sapwood of structural softwoods, causing severe destruction, even leading to structural collapse in some cases. The female beetle may lay up to 200 eggs and if most of these hatch within the same structure, very substantial damage may result before the adult beetles emerge. The life cycle of the insect varies from 3 to 11 years. Another woodboring species of major importance is the Lyctid family, the Powder Post Beetles. These infest recently seasoned hardwoods and feed only on the sapwood portion; softwoods are not attacked since the vessels of the wood are too small for the eggs to settle. The larval period is 1-2 years and the larvae can be identified by two brown spots, one on either side near the rear of the body; the Lyctidae are thin and are red-brown to black in colour. Although not particularly common in buildings, damage to strip or block flooring or to fitted furniture is expensive to eradicate. Of rather less importance are the wood weevils. These attack timber softened by fungal decay. Beetles, larvae and eggs can exist together in wood. The flight hole is irregular in shape and about

2.5 mm across. The life cycle of the weevil is about 1 year.

In tropical and semi-tropical regions the termite or "white ant" is a very
real problem, causing extensive damage to wood and fibre products. These termites bore into the wood, setting up nests and eradication has proved extremely
difficult.

4.5.2   Insecticidal treatments

Chemicals are used against wood-boring insects with the intention either of
protecting sound wood for the duration of its active life or of eradicating existing infestation in unprotected wood. The former is easier than the latter,
hence the emphasis on pretreating timbers for modern buildings. Tests to evaluate the effectiveness of biocides against these wood-borers fall into two categories: those where insects are introduced to wood which has been pretreated
(preservation) and those where an insecticide is introduced to wood already infested with wood-borers (eradication). Preservation tests involve confining
pairs of adult beetles on both treated and untreated blocks of wood. The effectiveness of the treatment in preventing infestation can then be assessed by comparison. In larval transfer tests, half-grown larvae are transferred to wood
samples; since these are less vulnerable than newly-hatched larvae when endeav-
ouring to bore into wood, this is a more stringent test of toxicity, but the
results are more difficult to assess in terms of real-life situations. Eradication of wood-boring insects is brought about by the combined effects of an initial kill, of killing survivors as they tunnel through treated wood and of protection against re-infestation. Tests have been devised to assess both initial
kill and long-term effect on survivors. It is not feasible to achieve complete
enough penetration of wood with toxicant to destroy larvae at a great distance
from the surface, but when returning to the surface, the pests die from contact
poisoning.

Tributyltin compounds are powerful contact insecticides but once they are
fixed in the wood they lose this direct contact action and it is necessary for
treated wood to be "tasted" before a lethal dose can be absorbed [28]. In the
case of Anobium punctatum and similar beetles in which attack is initiated by
egg-laying, the larvae rapidly assimilate a lethal dose and damage is invisible.
However, where attack is initiated by comparatively large insects such as termites, or in laboratory larval transfer tests, much more wood must be damaged
and digested before a lethal dose is imparted to the insect. During this period
the insect may be eliminating some of the toxicant and it is possible that a
lethal dose will never be reached unless higher treatment retentions are used.
Baker and Taylor [29] demonstrated that T.B.T.O. did not have any pronounced

contact effect on adult <u>Lyctus brunneus</u> beetles or on the larvae of <u>Anobium</u> <u>punctatum</u> in wood. Because of this limited effectiveness tributyltin compounds are often combined with powerful contact insecticides, particularly chlorinated hydrocarbons such as dieldrin and gamma benzene hexachloride. It has been shown that the fungicidal action of T.B.T.O. may be increased synergistically by the addition of these compounds.

Nicholas and de Ryte [30] have described the effectiveness of a combination of T.B.T.O. and dieldrin in preventing termite attack in both laboratory studies and field trials. In the field studies, treated wood samples were inserted in test chambers consisting of well-developed termite nests in wooden poles. Total exposure times (including exposure of the test blocks to several different colonies) ranged from 200 to 510 days. Of the treatments evaluated, the T.B.T.O./ dieldrin combination was by far the most effective and further series of tests covering 1225 days using six different termite colonies confirmed the effectiveness of the 0.5% dieldrin/0.5% T.B.T.O. treatment.

4.6 METHODS OF APPLICATION OF PRESERVATIVES

Organotin compounds are normally applied to wood in organic solvents (developments with water-soluble organotins are described in 4.8.1). The principal requirements are that the biocide should be absorbed by the wood at a sufficient concentration to provide protection, that it should achieve good penetration and that it should be distributed within the wood as uniformly as possible. This latter point is important since it has been seen that some fungi can pass through areas of high concentration to other less well-protected regions. Types of wood show differing abilities to take up preservative, for example, spruce is notoriously difficult to impregnate uniformly. The preservative quality and the treatment method must thus be matched to the type of timber and the hazard to which it is likely to be exposed. For example, a simple immersion treatment is sufficient for redwood in a low-hazard environment such as roofing timbers, whereas for wood exposed to continual wetting, a higher standard of treatment is needed.

4.6.1 Pretreatments for wood

The first preservative treatments consisted of brushing, simple dipping or spraying of the treatment solution. Whilst these are still appropriate in many cases, more sophisticated methods have been developed for pretreating joinery and structural timbers. The advent of the deluging machine was the first step forward in improving the processing operation for structural timbers. Manual dipping was thus replaced by a more mechanised and controllable treatment. When the need to process joinery arose, immersion of made-up frames or of fully mach-

Fig. 4.7. Timber is widely used in modern buildings and needs pretreatment with preservatives to ensure a long life free from decay.
Photograph courtesy: Fosroc Ltd., Marlow, Bucks., U.K.

ined parts before assembly for a period of 1 to 3 minutes was adopted. Certain specifiers from the outset required that a 3-minute timed controlled immersion be carried out and automatic plants for processing both joinery and carcassing timbers to these specifications were developed. Batches of timber are either lowered into an already-filled tank and locked in position for a pre-set time, or are placed in an empty chamber which is then flooded and drained after the required period. Mechanised immersion processes employ loading cages or cradles, preservative storage tanks, a transfer pump and an automatic timer for the treatment cycle.

A limitation of controlled immersion methods is that only the time of immersion is controlled, not the degree of absorption of the preservative. This latter depends on the species of wood, which may have ultra-absorbent zones and also on the dimensions of the individual wood pieces. Also, on removal from the dip tank, cells near the outer surface of the timber are flooded with preservative solution and the timber must be stacked after treatment so that it can drain and dry.

A significant development has been the introduction of controlled vacuum and pressure impregnation techniques. These enable a much more controlled uptake and distribution to be achieved in the wood and the treated product is dry to the touch on unloading and can be glued or painted within a very short time.

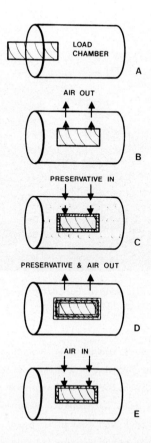

Fig. 4.8. Stages of operation of a double vacuum process for impregnation of wood.
A Load treatment vessel with wood to be treated.
B Adjust pressure of air trapped in the wood, by applying a vacuum.
C Flood treatment solution into the vessel and adjust vacuum to allow a predetermined excess of solution to be absorbed by the timber.
D Withdraw treatment solution and adjust vacuum to remove excess solution from wood.
E Release vacuum and remove timber from vessel.

These techniques are particularly suitable for pretreating, in a reproducible manner, joinery made to strict dimensional tolerances. The principle of operation is shown in Fig. 4.8. There are numerous commercial processes but the essential stages of double vacuum treatments are as shown.

By controlling the duration of the initial vacuum and the degree of vacuum in the various cycles, the absorption can be adjusted to suit a particular timber species and a required end use. These double vacuum processes require more

elaborate equipment than simple immersion methods and more preservative is applied to the wood (1.5-3 times as much) but there are significant savings in time, labour, space and stocks. Figs. 4.9 and 4.10 show typical commercial units.

In the case of certain timbers such as spruce, uptake and penetration are low
and to cope with these situations a double vacuum schedule with a pressurised
immersion phase has been introduced. When the pressure is maintained at 1 bar
above atmospheric this treatment is known as a "one-bar" schedule. Higher pres-
sures (up to 2 bar) facilitate a reduction in the immersion time. Saunders [31]
has examined the behaviour of organic solvent preservatives in European white-
wood during a one-bar schedule treatment. The influence of the degree and duration of the initial vacuum, applied pressure and final vacuum was studied and,
based on the findings, several mechanisms were proposed for penetration. The

Fig. 4.9. A modern "Vac-Vac" Plant for pretreating timber.
Photograph courtesy: Hickson's Timber Products Ltd., Castleton, W. Yorks., U.K.

first assumes that random, incomplete aspiration of pits results in clusters of
tracheids with interconnecting flow paths, but sealed off from surrounding timber. During immersion of spruce logs, many of these clusters are cut open so
that preservative gains easy access to the lumina of the tracheids in open clusters near the surface, but little penetration beyond. The second mechanism is
based on the fact that "white spirit" type solvents are very wetting to the timber. In Fig. 4.11, A and C represent lumina of the tracheids whilst B and D
represent pores in imperfectly aspirated pits. Liquid entering A will flow in,

128

Fig. 4.10. Removing treated timber from an impregnation plant.
Photograph courtesy: Fosroc Ltd., Marlow, Bucks., U.K.

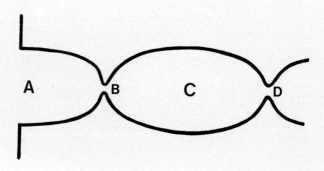

Fig. 4.11. Schematic model of a flow path for preservative in treated wood [31].

but as it advances towards C, its volume rate of advance will be retarded by the
narrowing flow path. As the liquid flows past B and into C, the maximum rate of
flow possible through B determines the subsequent behaviour. If the rate is
sufficient, void C will become filled before D. If the rate is insufficient,
liquid flowing into C may creep around the walls, not filling the void. Thus,
by the time the liquid reaches D, and subsequent constrictions, the walls of the
tracheid will be coated but the lumina will remain unfilled. The pressure dif-
ferential influences the rate of flow through B; a less severe initial vacuum

provides conditions for the second situation and a lower preservative retention in relation to penetration. An optimum treatment schedule related to the results, consisted of an initial vacuum of -0.17 bar for 10 minutes followed by a pressure of 1.0 bar for 40 minutes with a final vacuum of -0.85 bar for 30 minutes.

Jensen and Imsgard [32] examined the degree of axial penetration when window joinery is treated by the double vacuum technique, since this is of relevance to the protection of corner joints, where this timber is most susceptible to decay. The treatment was found to produce a continuous shield of preservative in all superficial layers of the wood, particularly in the cross-cut ends; axial water absorption was also considerably reduced. The authors recommended that double vacuum treatment and surface treatment should preferably be carried out by the window manufacturers before assembly of the frames, for maximum protection.

4.6.2 Remedial treatments

For treatment of affected wood in situ, clearly impregnation and immersion treatments are not practicable. In those cases brushes, spraying or injection techniques are adopted; in some instances special pastes and other techniques have been developed to ensure as deep a penetration as possible. Although preservative solutions are available commercially for domestic use, treatment of extensive damage is a job for the specialist. Consideration of the procedures involved in detail is outside the scope of this book and only a general outline can be given. Spraying is perhaps the most effective method of covering an affected area uniformly with organic solvent preservatives. Adequate precautions such as the use of face mask and goggles should be adopted (see 4.7). When spraying organic solvents in confined spaces the risk of fire must be taken into account. One commercial product, designed to overcome this hazard, contains the organic solvent/preservative system in an emulsion, each sprayed droplet having an outer film of water, thus reducing flammability. Paste formulations have also been developed which form a sealing skin when applied to the surface of the wood, the lower-viscosity preservative constituent migrating deep into the wood pores. These pastes are claimed to be effective in dealing with beetle-infested wood. With fungal attack, attacked timbers should be cleared and trimmed where necessary, in some cases structural repair being carried out. It should be remembered that in the case of some fungi, notably the dry rot fungus, hyphae may spread and penetrate over large areas so that extensive probing is necessary and affected masonry may have to be sterilised. In the case of insect attack, too, the presence of bore holes in one section may mean extensive internal damage over parts of the wood far removed from these holes. Preservative is usually injected into these holes under pressure so that all chambers are filled. The loading of biocide remaining at the surface ensures that any surviving larvae in the wood which pupate into beetles, will pick up sufficient contact poison on emerging to cause their subsequent death before any further egg laying can occur.

One rather novel method of applying wood preservative has been developed by a firm in The Netherlands [33]. The active fluid is contained in small sealed glass capsules which are inserted into bore holes or at locations specially vulnerable to decay such as joints or frameworks, and then broken by sealing the hole with a wooden plug. The escaping preservative then gradually diffuses into the wood structure.

4.7   PRECAUTIONS WHEN HANDLING ORGANOTIN PRESERVATIVES

All wood preservatives are necessarily toxic substances, but no harmful effects will be experienced in the use of organotin-based formulations providing manufacturers' instructions are followed. In the 25 years since their introduction, there have been no serious reported cases of health hazards arising from the use of these preservatives.

Gloves should be worn during handling of the preservative solutions and eyes should be protected by goggles. If the solution is splashed on the skin it should be washed off immediately with copious amounts of water; contaminated working clothes should be changed at once.

The U.K. Government's Health and Safety Executive operates a Pesticides Safety Precautions Scheme (PSPS) which has now been extended to cover wood preservative toxicants. Products are examined by an advisory committee which makes recommendations regarding their use. T.B.T.O. has been approved for use under this scheme in hydrocarbon solvents at levels of 3.0% for industrial pretreatment, at 1.5% for professional and 1.0% for domestic timber treatments. Tributyltin phosphate has also been approved in ready-for-use oil-based formulations at levels up to 1.5%.

With regard to pretreated timber, once the organic solvent has evaporated, the biocide is held firmly in the wood and treated wood presents no hazard in handling or sawing.

4.8   NEW DEVELOPMENTS IN ORGANOTIN PRESERVATIVES

Organotin compounds, particularly T.B.T.O., are now well-established in organic-solvent-borne wood preservatives and have a long record of success. Research is continuing not only into understanding the mechanism of protection and factors influencing it, but also into developing new compounds or new ways of utilising existing biocides.

4.8.1   Water-borne organotin preservatives

Water-based wood preservatives such as copper-chrome-arsenate (CCA) are widely used throughout the world and although organic-solvent-based systems are preferred for many applications, for example pretreatment of joinery with strict dimensional tolerances, the water-based treatments offer the advantages of being low-cost and free from fire risks in application.   To date, tributyltin compounds have not found much use in water-borne systems because of their low water solubility.   One approach has been to make T.B.T.O. water-dispersible by the addition of suitable quaternary ammonium compounds; these quaternaries are themselves active against gram negative organisms so that their incorporation also widens the range of effectiveness.   These formulations were originally developed for treatment of algal growths on masonry [34], but subsequently they were found to be also highly effective as wood preservatives.   In fact, the activity of T.B.T.O. against wood-destroying fungi appears to be enhanced in the presence of polar solvents and particularly in the presence of water.   It has been speculated that this may be due to the ability of water-borne tributyltin compounds to penetrate into the walls of the wood cells, thus protecting them from enzymatic attack; organic-solvent-borne organotins tend to be confined within the interior of the cells.   This aspect was discussed in 4.4.

A typical water-dispersed organotin concentrate contains 1 part T.B.T.O., 8 parts of a quaternary ammonium compound such as benzalkonium chloride (50% active agent) and 1 part water.   This overall 10% T.B.T.O. concentrate is then diluted with water to give any desired concentration of the organotin [35].   The first proprietary preservative to use this type of system was trade-named "Permapruf T".   Results of long-term trials have indicated that this is a viable alternative to water-based CCA formulations and is effective at lower retentions of active ingredient in sapwood [19].   Pressure impregnation treatment vessels have been modified so as to be suitable for impregnation of water with water-based organotins.   The formulation has been approved for use in Sweden (under the name BP Hylosan PT) by the Swedish Wood Preservation Institute [36].

Another approach, which has been followed in the laboratories of the International Tin Research Institute, has been to develop organotin biocides which are themselves water soluble.   Compounds of the type Bu_3SnSO_3R have been systematically studied to determine the effect of the R group on water solubility [37].   Of especial interest were the tributyltin alkanesulphonates which have aqueous solubilities in the range 0.7-1.5% w/v.   This would be high enough

for most biocidal applications. The water solubility of this group of compounds
has been found to increase with the electron releasing power of the R group,
with a maximum solubility for ethyl [38]. A further increase in chain length of
the n-alkyl group causes a drop in solubility. The biological activities of the
compounds do not appear to be reduced by their dissolution in water. Laboratory
evaluations of the fungicidal properties of tributyltin ethanesulphonate have
been conducted at the International Tin Research Institute using a modified wood
block/agar test to determine toxic limits and at Forintek Canada Corporation
using a standard soil block test to determine threshold limits. Performance was
seen to be comparable with that of T.B.T.O. in organic solvent [39]. Tests with
leached blocks suggest that tributyltin ethanesulphonate, though water-soluble,
is well fixed into the wood. An in situ ^{119m}Sn Mössbauer study on the compound
in Scots pine and Ponderosa pine sapwood suggested that the structure was virtually unchanged when impregnated into wood.

4.8.2   Wood preservation by organotin polymers

A novel approach to wood preservation has been adopted by Subramanian and co-
workers [40], who have attempted to produce an organotin polymer within the
wood. The active tributyltin moiety would then be released under controlled
conditions to protect the wood. A similar technique of controlled release of a
toxicant from a polymer has been adopted in antifouling technology (see Chapter
5). In practice, the wood is impregnated with an organotin vinyl monomer and a
co-monomer carrying a functional group capable of reacting with wood components.
These are then copolymerised by a charge transfer mechanism on heating. Typical
pairs of monomers were tributyltin methacrylate with maleic anhydride and tributyltin methacrylate with glycidyl methacrylate. These were dissolved in benzene
in the presence of a free radical initiator such as benzoyl peroxide, impregnated into wood samples and heated to initiate vinyl polymerisation and functional
group/wood hydroxyl reactions. Grafting of polymer to the wood was high. The
macrodistribution of tributyltin copolymers in the treated wood declined from
the surface to the centre; electron microprobe analysis for tin atoms showed
that a significant amount of the polymer was located in the cell walls.

In a second paper [41] the strength and decay resistance of treated wood samples was investigated. In the transverse direction the strength of the wood was
increased significantly by the treatment but only moderately so in the longitudinal direction. The thickness swelling and water absorption of the treated wood
were substantially less than for untreated wood. Specimens were tested against
brown rot, white rot and soft rot fungi as well as against a marine bacterium,
and showed effective protection against biodeterioration. These studies were
considered to have shown the in situ polymerisation technique to be capable of

improving the mechanical and decay-resistant properties of wood.

4.9   THE FUTURE FOR ORGANOTIN WOOD PRESERVATIVES

Organotin compounds have played a significant part in the evolution of pre-treatment techniques for building timbers. Timber can be pretreated with guaranteed protection and with no undesirable side effects. The mechanisms by which organotin compounds exert their protective action and the ways in which the compounds are held in the wood are the subject of continuing studies and a greater understanding of these processes will help in making wood preservation even more effective.

Another important line of study consists of determining the extent to which tributyltin compounds degrade in the wood and the possible effects on long-term preservation. In the field of new biocides, compounds are being developed which may increase the potential of organotins as wood preservatives; one example is the water-soluble compound now being tested. In many formulations chlorinated hydrocarbons are used in conjunction with organotins and since legislation is restricting the use of these compounds in many countries, it will be necessary to find new partners for the organotin. One possibility is the use of synthetic pyrethroids. There is, of course, always the possibility of developing an organotin biocide which would have the necessary spectrum of biocidal activity to be used alone. All in all, there would appear to be an exciting future for organotin-based wood preservatives.

REFERENCES

1   A.C. Oliver, Timber Trades J., June 26th (1971) 4-5, 7.
2   G.J.M. van der Kerk and J.G.A. Luijten, J. Appl. Chem., 4 (1954) 314-319.
3   C.J. Evans, Tin and its Uses, 115 (1978) 11-13.
4   U. Thust, Tin and its Uses, 122 (1979) 3-5.
5   Anon., Organotin Compounds : Wood Protection, Schering AG Publicn., 1983, 24 pp.
6   E.A. Hilditch and R.E. Hambling, Brit. Wood Preserv. Assn. Ann. Conven., July, 1971, pp. 95-113.
7   A.J. Crowe, R. Hill, P.J. Smith and T.R.G. Cox, Int. J. Wood Preserv., 1 (1979) 119-124.
8   B.A. Richardson, Wood Preservation, Construction Press, 1978, 238 pp.
9   D.J. Dickinson and J.F. Levy, Brit. Wood Preserv. Assn. Ann. Conven., June, 1979, pp. 33-37.
10 A.F. Bravery, Proc. Sessions on Wood Products Pathology, of 2nd Internat. Congress of Plant Pathology, Sept. 10-12th, 1973, Minneapolis, pp. 129-142.
11 N.E. Hicken, The Rentokil Library - The Dry Rot Problem, Hutchinson, London, 2nd Ed., 1972, 200 pp.
12 D.J. Dickinson, Int. Pest Control, 14 (1972) 21-25.
13 B.A. Richardson, Brit. Wood Preserv. Assn. Ann. Conven., July, 1972, pp. 613-614, 617-620, 623.
14 B.A. Richardson, Proc. 1st Int. Biodet. Symp., 1968, pp. 498-505.
15 P.J. Smith, A.J. Crowe, D.W. Allen, J.S. Brooks and R. Formstone, Chem. Ind., 5th Nov., (1977) 874-875.

16 B.A. Richardson and G.H. Lindner, 2nd Penarth Conf., June, 1979, 12 pp.
17 W.N. Aldridge, Am. Chem. Soc. Advanc. Chem. Ser., 157 (1976) 186-196.
18 A.F. Bravery, Proc. Conf. Int. Symp., Berlin, 1975; Mater. u. Organismen Chem. Supp., 3   331-344.
19 B.A. Richardson and T.R.G. Cox, Tin and its Uses, 102 (1974) 6-10.
20 A.F. Bravery, Int. Biodet. Bull., 6 (1979) 145-147.
21 T.R.G. Cox, Int. J. Wood Preserv., 1 (1979) 173-176.
22 C.R. Coggins, Brit. Wood. Preserv. Assn. Ann. Conven., June, 1982, pp. 65-71.
23 J. Jermer, M.L. Edlund, B. Henningson, W. Hintze and S. Ohlsson, Paper to 14th Ann. Meeting Int. Res. Gp. Wood Preserv., 9th-18th May, 1983, Australia, Doc. IRG/WP/3219.
24 J. Jermer, M.L. Edlund, W. Hintze and S. Ohlsson, Paper to 14th Ann. Meeting Int. Res. Gp. Wood Preserv., 9th-18th May, 1983, Australia, Doc. IRG/WP/3230.
25 H. Plum and H. Landsiedel, Holz Roh-Werkst., 38 (1980) 461-465.
26 Anon., Tin and its Uses, 138 (1983) 12-13.
27 C.J. Evans and R. Hill, J. Oil Col. Chem. Assn., 64 (1981) 215-223.
28 B.A. Richardson, Brit. Wood Preserv. Assn. Ann. Conven., June, 1970, pp. 19-26.
29 J.M. Baker and J.M. Taylor, Ann. Appl. Biol., 60 (1967) 181-190.
30 D. Nicholas and R. de Ryte, Brit. Wood Preserv. Assn. Ann. Conven., July, 1972, pp. 129-139.
31 L.D.A. Saunders, Brit. Wood Preserv. Assn. Ann. Conven., June, 1982, pp. 46-54.
32 B. Jensen and F. Imsgard, ibid., pp. 55-64.
33 Anon., "Woodcap", Tech. Note Hoeka Sier pleisters en Muorverven B.V., Deurne, The Netherlands.
34 B.A. Richardson, Stone Ind., 8 (1973) 2-6.
35 Anon., TBTO in Water-Based Preservatives for Wood, Albright and Wilson Ltd., Tech. Service Note TS/RECH/C5, 1975.
36 S. King, Tin and its Uses, 124 (1980) 13-16
37 S.J. Blunden, A.H. Chapman, A.J. Crowe and P.J. Smith, J. Intern. Pest Control, 20 (1978) 5, 8, 12.
38 A.J. Crowe, R. Hill and P.J. Smith, Tributyltin Wood Preservatives, Int. Tin Res. Inst. Publicn. 559, 7 pp.
39 R. Hill, P.J. Smith, J.N.R. Ruddick and K.W. Sweatman, Paper to 14th Ann. Meeting Int. Res. Gp. Wood Preserv., 9th-18th May, 1983, Australia, Doc. IRG/WP/3229.
40 R.V. Subramanian, J.A. Mendoza and B.K. Garg, Holzforschung, 35 (1981) 253-259.
41 R.V. Subramanian, J.A. Mendoza and B.K. Garg, ibid., 263-272.

Chapter 5

ANTIFOULING SYSTEMS

5.1   INTRODUCTION

Water is the natural milieu for a wide range of living organisms; it has been estimated, for example, that one millilitre of seawater contains between 1000 and 10,000 living cells. Many of these are slime-forming bacteria, others are plant forms such as algae or marine animals such as hydroids, crustaceans, molluscs and tunicates. Any surface which is immersed in water for any length of time becomes part of this environment and a target for attachment of many of these aquatic organisms; this includes the hulls of vessels. Marine fouling can be a serious problem in the shipping industry, since it increases the surface roughness of the hull and hence its frictional resistance to movement through water. This results in a loss of speed or increased fuel consumption to maintain the same speed. The frequency of dry-docking is also increased and re-coating incurs additional expense. Fouling also hampers inspection and maintenance and may accelerate localised corrosion. It has been estimated that a loss of 0.1 knots on a service speed of 16 knots for a supertanker can increase running costs by as much as 0.5%. About 25% of increased fuel costs have been attributed to marine fouling [1]. It is thus clear that fouling presents a serious economic limitation. Again, on static offshore structures such as oil and gas platforms, marine organisms can increase the loading forces caused by waves and currents, possibly even exceeding design specifications.

Various methods of preventing marine fouling have been adopted down the ages, one of the oldest employing copper sheathing on wooden ships (copper ions have a biocidal action). It was reported by Atheneus in 2000 BC that: "The ships of Archimedes were fastened everywhere with copper bolts and the entire bottom was sheathed with lead". Arsenic compounds and sulphur were formulated in wax- and tar-type coatings to protect ships' bottoms as early as 300 BC [2]. The first industrial use of an antifouling paint seems to have been in 1860, when a soap based on copper sulphate was applied hot on a primary layer of colophony and iron oxide. Since then, paint formulations have developed into complex systems and are the subject of continuing research.

The basic principle of an antifouling coating is that the toxic agent should be released slowly at a constant rate by the coating in contact with water, so that a thin boundary layer rich in toxicant is produced which will kill or at

least repel marine organisms approaching the coating. The thickness of this
boundary layer depends on the temperature, pH and salinity of the water and the
rate of flow of water over the surface. The rate at which the toxicant is
leached from the coating is critical : too slow a rate and protection is incomplete, too fast and the life of the coating is reduced and a risk of pollution
is introduced since fouling can be caused by a wide range of organisms, it is
also important that the toxicant has a wide range of biocidal activity. Tributyltin and triphenyltin compounds are establishing a secure position as the active ingredients of antifouling coatings, some 2000-3000 tonnes being consumed
each year by this application.

   Since marine fouling is a complex process, varying from one geographical region to another, so the type of antifouling coating applied must be tailored to
suit the operating conditions. Trading patterns may be normal (long sailing
periods, short stays in port) or severe (frequent or long periods in port with
heavy fouling conditions at sea). The operating speed of the vessel, its hull
surface roughness and dry-docking schedule must all be taken into account. The
types of fouling prevalent in normal trading waters must also have a significant
bearing on the type of protection needed. As will be seen, coating systems have
been developed which meet most of the requirements of the marine industry as
well as for the home boat owner.

5.2   PROBLEMS OF MARINE FOULING

   The range of fouling species includes animal and plant forms and the type
predominating depends on the location and the season. They are essentially sed-
entary organisms which establish themselves on substrates supplying a fairly secure base. Once settled, they can then complete their life history. The principal types of fouling can be characterised under various headings.

5.2.1   Primary film

   The term "primary film" is normally used to describe the complex community of
bacteria, diatoms, protozoa and algal spores, together with organic and inorganic detritus, which become attached to surfaces immersed in water. Corpe [3] has
demonstrated that when an unprotected surface is immersed in the sea, it quickly
absorbs organic compounds and within hours, a flourishing growth of bacteria develops. This will be followed by the microscopic algae, the diatoms and those
animals which display plant-like characteristics, which together constitute the
so-called "slimes". As antifouling formulations have begun effectively to combat the animal and plant types of fouling, so the more resistant slimes are as-
suming a more important role. Particularly with slow moving ships, the increase
in frictional resistance due to slimes becomes more of a problem and may account

for 5% of total frictional resistance. The major components of slime are cat-
egorised in Table 5.1 [1].

TABLE 5.1

Major components of slime

Organisms	Bacteria (0.1 μm)	Pseudomonas sp. Achromobacter sp. Flavobacterium sp. Caulobacter sp. Saprospina sp. Micrococcus sp.
	Diatoms (5-100 μm)	Acnanthes sp. Nitzchia sp. Navicula sp. Diploneis sp. Mestogloia sp. Amphora sp.
Non-organisms	Organic	Dead plankton, excrement of fish
	Inorganic	Floating minute particles

A feature of fouling by slimes is its cyclic nature. When vessels are stationary, the slime film builds up and is partially wiped off during hull motion,
only to build up again in the next stationary period. Certain thiocarbamate
compounds have been found highly effective against slimes when used in conjunction with tributyltin compounds.

5.2.2   Principal fouling organisms

The major fouling forms include barnacles and other molluscs, ship-worms and
tube-worms and algae, particularly the sea grasses, Enteromorpha sp. Barnacles
(Cirripedia) are probably the most commonly observed macrofouling organisms on
submerged structures. There are two main types which are characterised by the
shape of their shells. The first is the "acorn barnacle" which has a truncated
conical calcareous shell; the second is the stalked or "goose barnacle" which is
attached to a substratum by a fleshy stalk and which carries the shell portion
at its tip. Barnacles are a major problem for vessels in port; the larvae are
free-swimming and become attached to an immersed surface within 24 hours by se-
creting tiny drops of a cement. Interestingly, this cement adheres to all tested substrates including glass and p.t.f.e. The larvae then grow into the conical adult form with calcareous shells. Barnacles do not require sunlight for
survival and thus are able to become firmly attached to the bottom of a vessel.

Detachment of the acorn barnacle is generally more difficult than removal of other fouling organisms. With a density close to that of calcium carbonate, they add true weight to the immersed vessel, whereas non-shelled foulants vary little in weight from seawater.

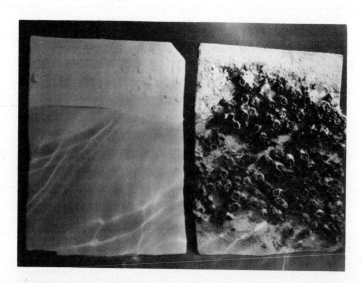

Fig. 5.1. The panel on the right shows a typical incrustation of acorn barnacles; the panel on the left was protected with an antifouling paint containing organotin.
Photograph courtesy: Schering AG, Bergkamen, F. R. Germany.

Oysters and mussels are other molluscs which can constitute a source of fouling; these are bivalves, i.e. they have two calcareous shells or "valves" enclosing the soft-bodied animal. Tube-worms (Serpulida) are characterised by a calcareous tube which attaches itself firmly to the substratum during settlement. They may be 7 cm or more in length and 2 cm in diameter at the distal end. The most important and frequently observed species in this family is Hydroides norvegica; another family within the class Polychaeta are the Spionids sp. These worms also live permanently within a tube, but in this case it is composed of mud and silt particles held together by a type of mucus secreted by the worm. Ship-worms which can attack wooden structures include Teredo worms which burrow into wood and Limnoria which actually ingest wood. It is interesting to note that the devastating effect of these foulants on wooden-hulled ships is given more credit than Sir Francis Drake for the destruction of the Spanish Armada [4].

Algae are commonly referred to, as "seaweeds" or "kelp" and include a large number of microfouling species already discussed, typified by the diatoms. Algae are divided into groups on the basis of their pigmentation, namely blue-green algae, greens, reds and browns. Since they are plant forms, algae require daylight for growth, so that they are restricted to the vicinity of the water-line. One of the most common species of algae to be found in ship fouling are the Enteromorpha sp. (green ribbon grass) [5]. The dominance of this species in marine fouling has been ascribed to its ability to withstand wide fluctuations in environmental conditions (for example water salinity and temperature) plus its rapid and highly effective method of spore attachment in response to surface contact (thigmotaxis). Christie [6] has described the mode of attachment of this organism; initiation of growth is by means of flagellated spores, some 10 μm long, which can attach themselves to a roughened surface within seconds, even when the surface is moving at up to 10 knots [7]. Attachment is made by means of a glycoprotein adhesive secreted in vesicles within the spore. In a few hours, following attachment, a relatively thick protective cell wall is formed, which thickens and grows, the only requirements being a supply of light and nu-trients. Growth may be as much as 15 to 25 cm within a few weeks. Other sea grasses which present problems include the brown algae Ectocarpus sp. and the small filamentous green algae Ulothrix sp. These both grow as dense mats; Ectocarpus has a horizontal creeping system of filaments from which erect filaments arise.

With fast, quick-turn-around container ships and tankers, algal settlement between the deep and light load lines is the major fouling situation [8]. Over the large surface area of a tanker the frictional drag and resultant increased fuel costs can be considerable. A survey of fouling on 20 oil tankers docking in European waters showed the problem to be nearly all algal in nature, with a limited number of species broken down as follows [9]:

Enteromorpha sp.	75%
Ectocarpus sp.	13%
Chladophora sp.	5%
Chaetomorpha sp.	5%
Others	2%

Vessel speeds in excess of 2 knots would effectively prevent settlement of animal species, but in situations involving extensive stays in ports, or in the case of stationary structures these forms can readily attach themselves to the surface. In these circumstances, Bryozoans are one of the most common groups of marine fouling organisms and comprise a great variety of species possessing widely different forms and colonies. Two basic forms are readily distinguish-able; the first is represented by the Bugulae sp. which are plant-like, erect and branching. The second is exemplified by <u>Watersipora sp</u>. and <u>Cryptosula sp</u>. which form a flat, encrusting, often brittle calcareous lacework over the sub-stratum. Sponges (<u>Porifera</u>) are not generally found to any great extent on ships' hulls. However, under certain conditions, they can constitute a large percentage of the fouling biomass on pier piles. Sea squirts (Ascidians) are a group of soft-bodied animals protected by an outer skin or test. Some species occur as solitary individuals living independently, whilst others form colonies with the small individuals more or less embedded in a common matrix, often of a gelatinous consistency. Hydroids are colonising animals which frequently present a plant-like appearance. Comparatively few of the species are found as fouling organisms. A more detailed description of marine fouling organisms can be found in a series of O.E.C.D. Catalogues which illustrate the various classes [10].

5.2.3   Effect of environmental conditions

Fig. 5.2 shows the rate of fouling of an unprotected surface in the North Sea. Fouling is greatest near the surface where warmth, sunlight and nourishment are more readily available.

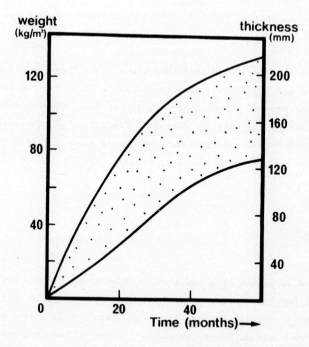

Fig. 5.2. Development of weight and thickness of a fouling layer with time, in the North Sea [1].

In temperate waters, fouling is seasonal, with organisms which are practically inactive during the winter, growing rapidly between April and September and
liberating spores and larvae. In tropical waters on the other hand, organisms
grow all year round, so that fouling is a greater problem. It has been calculated that a 10°C rise in water temperature doubles the rate of growth for many
organisms. In a fouling population, each organism has a part to play in decid-
ing the rate of growth of the community and its constitution.

5.3   MODE OF ACTION OF ORGANOTINS

As early as 1943, a patent taken out by Tisdale [11] had claimed that organotin compounds were effective for antifouling systems. However, this patent
aroused no commercial interest and it was not until the 1960s that organotins
began to find use in antifouling paints, initially as a "booster" for cuprous
oxide-based systems and then in vinyl and acrylic paint systems as the sole toxicant. The early developments have been reviewed by Bennett and Zedler [12].

Bis(tributyltin) oxide (T.B.T.O.) was one of the first organotins to be employed in these coatings. The organotin is a liquid and is used mainly in paint
systems where relatively short-term protection is needed, since its leaching
rate from a paint film is high. Tributyltin fluoride is a white solid and is
used in longer-life antifouling paints, as are triphenyltin fluoride, chloride
and hydroxide. Tripropyltin compounds are excellent fungicides and broad spectrum bactericides and have been used in combination with tributyltin compounds
to broaden their range of activity. The solubility of the organotin compound in
water is important and differences in solubility require variations in formulation of the coating system to allow sufficient toxicant to be present in the
water layer surrounding the painted hull. Solubilities in seawater for typical
compounds are: 8-10 ppm (T.B.T.O.), 6 ppm (tributyltin fluoride) and 1 ppm (triphenyltin fluoride). The comparative antifouling activity of triorganotin compounds in relation to other biocides is shown in Table 5.2.

The exact mechanism by which triorganotins exert their lethal effects on marine organisms is still not completely clear, although extensive studies on mammals [14, 15, 16] have established that the lower trialkyltin compounds are able
to inhibit the vital process of oxidative phosphorylation and therefore disrupt
the fundamental energy processes in living systems. Hall and Zuckerman [17]
have suggested that the biological activity pattern of triorganotin compounds
may be dependent upon the effectiveness of their interaction at an active site,
or at sites which involve co-ordination to certain amino acids. In vitro studies with T.B.T.O. in snail protein indicated a high order of activity with a
number of amino acids (as well as lipids) and Cardarelli [18] has suggested that

mortality arises from direct reaction of the organotin with proteins. Millner and Evans [19] studied the effects of triphenyltin chloride on respiration and photosynthesis in the green algae Enteromorpha intestinalis and Ulothrix pseudoflacca. In both cases, zoospores of these species showed greater sensitivity to inhibition than did vegetative tissue. Results using vegetative Enteromorpha intestinalis also suggested that triphenyltin chloride acts as an energy-transfer inhibitor in vivo for both respiration and photosynthesis. However, a study by Boulton et al. [20] on the effects of organometallic biocides, including triethyltin sulphate, on the barnacle Elminus modestus presented a rather obscure picture of the toxic action. The organotin did not inhibit any of the enzymes sufficiently strongly to account for its toxicity, neither did its action as an uncoupling agent appear to correlate with its antifouling properties. It was considered possible that the metabolic effects of the organotin, which imply a restriction of pyruvate utilisation, resulted from its action on the dihydrolipoyl trans-acetylase component of pyruvate dehydrogenase. Table 5.3 shows the concentrations of T.B.T.O. and tributyltin fluoride needed to inhibit growth of various marine species [21].

TABLE 5.2

Comparative antifouling activity of various toxicants [13]

| Compound | LC_{50}, ppm | |
	Algae	Barnacles
Cu_2O	1-50	1-10
R_3SnX	0.01-5	0.1-1
* R_3PbX	0.1-1	0.1-1
* RHgX	0.1-1	0.1-1

* Because of the long-term environmental hazard of lead and mercury compounds, these biocides are not permitted for use in antifouling coatings in many countries.

TABLE 5.3

Growth inhibition by organotins of marine fouling organisms

| Organism | Growth-inhibiting concentrations, ppm | |
	T.B.T.O.	Tributyltin fluoride
Barnacles	0.1	0.1
Enteromorpha sp.	0.02	0.01
Chlamydomas sp. (algae)	0.005	0.001
Lobsters	0.02	0.005

In the case of enteromorpha, the resistance to toxicants increases with time, since the filaments, once consolidated, grow outwards at a rapid rate until they have passed the boundary layer of toxicant. Christie [9] found that spores of enteromorpha exposed to triphenyltin chloride five minutes after settling, gave an LD_{50} value of 0.002 ppm whereas when toxicant was added 60 minutes after settling, the LD_{50} value had risen to 0.005 ppm.

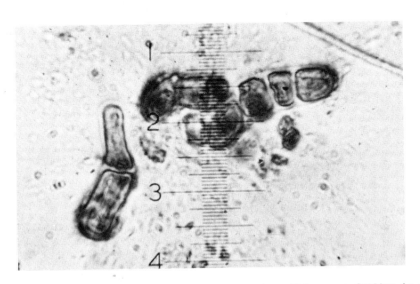

Fig. 5.3. Three-day old sporelings of Enteromorpha. This grass fouling is capable of rapid growth and an average 100 mm long filament can yield 4 thousand million reproductive spores.
Photograph courtesy: International Paint Marine Coatings, Gateshead, U.K.

5.4   THE SLOW RELEASE CONCEPT

5.4.1   Conventional paint systems

For an antifouling coating to be effective, the rate of release of the toxicant must be closely controlled. It must be great enough to provide a protective barrier, but not too great, or else the life of the coating is shortened and environmental contamination becomes unacceptable. The critical leaching rate for copper-based toxicants is about 10 $\mu g/cm^2/day$ [7] and it has been estimated that around 5000 tonnes of copper is released into the sea every year from copper-based antifouling systems [13]. In the case of organotin compounds these are far more potent biocides and the critical leaching rate for effective control is an order of magnitude lower.

A paint has three essential elements: a binding polymer, a solvent or "vehicle" and various pigments that provide colour, improved endurance, etc.

Antifouling additives confer biological activity. Various additives such as plasticisers, drying oils and anti-oxidants are used to enhance desirable chemical or physical properties and to eliminate undesirable ones.

The first commercial antifouling paints based on organotins employed T.B.T.O. as toxicant. Being a liquid, it was thought that diffusion to the paint surface would be easy. However, leaching rates proved to be too fast for most applications, where a long protective action was required, and its use is now mainly confined to coatings for the yacht and dinghy trade where protective action is only needed for a season. Solid organotins such as tributyltin fluoride and the triphenyltins offered more potential for controlled leaching rates and most subsequent work was done with these compounds. Conventional vinyl- or acrylic-based paints were used to contain the toxicants and a degree of formulation proved necessary to achieve the desired release rates. Mearns [22] has looked closely at the factors influencing the formulation of vinyl-based systems. High performance and long life must be based upon a non-saponifiable resin system which will provide a skeleton or sound framework in the exhausted film. The toxicant migrates to the film surface, stays there for a finite, if short, time and then transfers across the surface into the water phase. It exists as a particular concentration in the water around the surface dispersing continuously but with a defined half-life for a specified concentration. The relative rates at which these steps occur depend on a number of factors, for example whether the vessel is moving or stationary. The layer of toxicant in the water may be very thin; it has been estimated at no more than 0.07 mm at a ship speed of 1/4 mph. Outside this toxicant layer the amount of organotin reaching the sea has been said to be several orders of magnitude less than one millionth of a ppm, which would pose no toxicity hazard for fish or other marine life. The following factors are stated to influence leaching performance [2]:

Total percentage solubles
Paint resin/rosin ratio
Level of toxicant
Solubility of toxicant in seawater
Film thickness
Type of resin
Compatibility of toxicant with film components
Pigment volume concentration
Solvent characteristics
Presence of other additives (e.g. zinc oxide)
Adsorption of toxicants on pigments

Rosin is often added to antifouling paint formulations since it has a significant influence on film permeability and hence on the toxicant release rate. On
the other hand, water sensitive substances like zinc oxide will act as retardants. Differences in solubility of the organotin biocides will necessitate a
corresponding adjustment of the resin/rosin ratio and the level of zinc oxide.
Tributyltin fluoride gives a low release rate and may be used in rosin-contain-
ing paints (see Table 5.4).

However, Ghanem [23, 24] has studied the effects of graded amounts of rosin
and different types of pigment extender (varying in particle size, shape and
texture) on the antifouling performance of triphenyltin fluoride in conventional
paint systems and found, unexpectedly, that rosin had a deleterious effect,
leading to the formation of insoluble calcium soaps which hindered biocide release. China clay was the most effective extender, contributing to efficient
release and helping to maintain a clean surface for long periods.

In triorganotin compounds R_3SnX, the nature of the group X will affect the
solubility and hence the release rate. Most of the effective triorganotin antifoulants (with the exception of T.B.T.O. and triphenyltin chloride, which are
simple monomeric species) are self-associated polymeric molecules, due to intermolecular co-ordination by the X groups to tin. Examples are tributyltin and
triphenyltin fluoride (Fig. 5.4a) and tributyltin acetate (Fig. 5.4b).

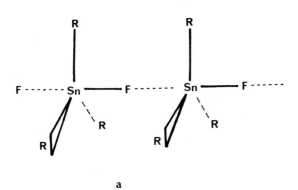

a

Fig. 5.4a. Solid-state structure of triorganotin fluorides.

Leaching tests with coatings made up using various organotin biocides with
varying amounts of binder, rosin and zinc oxide, correlated with raft tests to
determine antifouling efficiency, showed that greatest effectiveness was

TABLE 5.4

Influence of rosin on the leaching rate of tributyltin fluoride in vinyl ship-bottom paints exposed in Biscayne Bay, Miami, FL, U.S.A. [2]

| Exposure Months | Vinyl/rosin ratio | | | | | | | | |
| | 3 : 1 | | | 1.5 : 1 | | | 1 : 2 | | |
	% Residual	% Loss	F.R. *	% Residual	% Loss	F.R. *	% Residual	% Loss	F.R. *
0	100	–	–	100	–	–	100	–	–
3	82.5	17.0	100	37.5	63.5	100	18.9	81.1	100
6	71.0	11.5	100	18.5	19.0	100	10.2	8.7	35
10	70.0	1.0	90	18.3	0.2	80	3.5	6.7	0
14	64.3	9.2	90	–	–	55	–	–	0
16	67	–	85	–	–	40	–	–	0

* F.R. = Fouling Ratio.

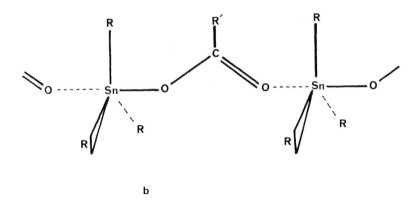

b

Fig. 5.4b. Solid-state structure of triorganotin carboxylates.

achieved with tributyltin and triphenyltin fluorides, a leaching rate of 1-2 µg/cm^2/day being sufficient to prevent fouling [25]. The nature of the plasticiser used in the paint film and the type of pigment, also have an effect on the leaching rate. Lorenz [26] has compared the effectiveness of T.B.T.O. and tributyltin tetrachlorophthalate in vinyl and chlorinated rubber coatings. The tetrachlorophthalate behaves as a plasticiser and the compound gave a lower leaching rate and greater long term effectiveness. However, Ghanem [23] found that plasticisers did not exert any significant effect on the leaching rate of tributyltin fluoride from vinyl paints.

Kronstein [27] has examined the effect of pigment loading on the leaching rate of organotin biocides from an alkyd resin/polyvinyl acetate latex paint system (Table 5.5). Since the initial pigment loading is much greater than the level of toxicant, the pigment leaching rate is naturally greater, but it is interesting to observe that in the paint formulation the leaching rate of toxicant is influenced by the nature of the pigment, and that for a given pigment, the nature of the organotin has a significant effect.

The incorporation of 0.1-10 wt.% of colloidal silica in a paint formulation containing tributyltin oxide has been patented [28] as a means of overcoming possible incompatibility of the organotin with the paint binder. Advantages claimed include better adhesion of the paint film and a slower leaching rate of the toxicant.

Beiter [29] showed that improved antifouling performance could be obtained when a combination of tributyltin fluoride (for early protection) and triphenyl-

TABLE 5.5

Pigment and biocide leaching from paint films

Nature of pigment	Nature of toxicant	Time (days)	Leaching rate $\mu g/cm^2/day$	
			Pigment	Toxicant
Zinc oxide	Tripropyltin acetate	15	2.97	0.49
		14	1.37	0.08
Zinc sulphide	Tripropyltin acetate	30	1.40	0.06
Iron oxide	Tributyltin fluoride	21	0	0.06
		34	0	0.07
		57	0.74	0
		65	0.65	0
Iron oxide	Tributyltin methacrylate	21	0	0.12
		35	0	0.14
		57	0.35	0
		65	0.31	0

tin fluoride (for long-term protection) was used in a hydrolysable carboxylated polyvinyl acetate paint system.

A number of leaching mechanisms have been proposed for paint systems. Monaghan et al. [30] suggested that a diffusion model proposed for leaching of cuprous oxide from vinyl coatings could also be used to describe the leaching behaviour of R_3SnX compounds (Fig. 5.5). R_3SnOH is considered to be the species actually leaching out of the coating. In the case of a vinyl and vinyl/rosin coating containing T.B.T.O., the organotin reacts with the resin to form tributyltin carboxylates, but this does not happen with other trialkyltin compounds.

Good et al. [31] have also proposed a model to explain the mechanism of toxicant release. They derived an empirical expression for leaching, based on Fick's First Law of diffusion and were able to show that this could be used to fit leaching data for tributyltin compounds such as the fluoride, acetate or oxide and for triphenyltin chloride or acetate, from conventional coatings.

Recent work has adopted a rather different approach to antifouling paint coatings, in an attempt to extend their active life. Microencapsulates of organotin biocide have been prepared and incorporated in paint films acting as reservoirs of toxicant. To make the encapsulate, tributyltin chloride, macro-

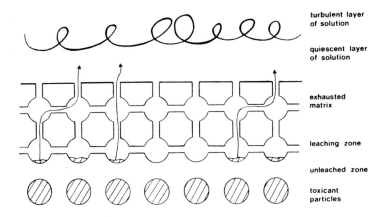

	turbulent layer of solution
	quiescent layer of solution
	exhausted matrix
	leaching zone
	unleached zone
	toxicant particles

Fig. 5.5. Diffusion model for leaching of toxicant from a paint film, according to F. Marson (J. Appl. Chem. 19 (1969) 93).

emulsified in water, was added to gelatin Type A and hydrated acacia, at 55°C. As the temperature was reduced, the two hydrophilic colloids formed coacervates around tiny microscopic droplets of the water-immiscible organotin. When cooled to 10°C, the encapsulated organotin was washed and the capsule walls fixed with glutaraldehyde. The product was then washed and dried to produce a loose powder consisting of tributyltin chloride coated with a dry macromolecular membrane. These capsules were incorporated at 14 wt.% in vinyl/rosin base paints and coated panels were tested in the fouling waters of Biscayne Bay, FL, U.S.A. After 4 years, test panels were still free from fouling.

5.4.2 Elastomeric systems

Another approach to antifouling coatings has been to incorporate triorganotin toxicants in an elastomeric matrix. Development in this field, leading to a commercial product, has been fully described by Cardarelli [32]. An elastomer is basically a polymeric material that will stretch under tension and, when the force is released, will return to its original shape and size. Of significance for controlled release systems, however, is the fact that from the standpoint of thermodynamics and kinetics, elastomers function like liquids at room temperature. Thus, when considering a toxicant dissolved or dispersed in an elastomer, we have a solid-in-liquid or a liquid-in-liquid solution or suspension situation. In order to change the elastomer from its original "gummy" state into a useful material, it is necessary to cross-link the smaller macromolecules to

create extremely large ones. This is achieved by incorporating a cross-linking
agent and inducing vulcanisation by heating. When tributyltin oxide is incorporated in chloroprene rubber, the organotin is thought to play a part in the
vulcanising process and the active agent is only partly tributyltin oxide, the
main constituents being tributyltin chloride and various tributyltin-fatty acid
combinations resulting from the vulcanising reaction [33]. The leaching mechanism is a diffusion-dissolution process. Organotins have a specific small solubility in elastomers and once incorporated at a concentration below, or at, the
solubility limit, a solution equilibrium exists.

Fig. 5.6. Sonar dome made of antifouling rubber containing 6.8% T.B.T.O.
Photograph courtesy: B. F. Goodrich, Akron, OH, U.S.A.

As T.B.T.O. dissolves from the surface, a localised solution disequilibrium
is set up and internal solute molecules migrate towards the depleted area so
that equilibrium is re-established. The net result of this continuous process
is that there is a constant supply of toxicant at the surface and no pore structure is set up in the coating. Release rates can be adjusted by selection of
specific polymers, compounding ingredients, vulcanisation conditions, coating
thickness and toxicant loading. These impregnated elastomers are applied to
surfaces in sheet form and provide the added bonus of a corrosion-resistant barrier layer.

One of the first commercial products of this type was trade-named "Nofoul"
and manufactured by B. F. Goodrich of Akron, OH in the U.S.A. Nofoul comprises

chloroprene rubber containing T.B.T.O. and tributyltin sulphide. Antifouling effectiveness in tropical waters has been displayed for over 84 months [34]. The first practical use of "Nofoul" was to protect sonar buoys (see Fig. 5.5) which play a vital part in the U.S. Navy's anti-submarine devices; it has since been extended to other applications. Woodford et al. [35, 36] have reported results of work on similar systems which has been conducted at the Australian Defence Standards Laboratories. The compounds studied were tributyltin oxide, acetate and fluoride at the 2, 5 and 10 phr level in natural rubber, nitrile rubber and polychloroprene containing conventional curing agents, antioxidants and carbon black. Effective antifouling protection was observed with all the systems studied, over 12 months immersion at Port Jackson, N.S.W. Tributyltin fluoride was considered the most promising toxicant for further work, on account of its low solubility in both water and elastomer. The degree of dispersion of the organotin within the rubber matrix has been found to be important and for this reason the less mobile tributyltin acetate proved itself less effective than the other organotins.

When organotins are incorporated in elastomers, the curing may be retarded and toxicant levels must therefore be restricted to 10% or less. Dunn and Oldfield [37] overcame this limitation by producing vulcanisates containing active organotin groups; this was achieved by adding an unsaturated organotin monomer, such as tributyltin acrylate to an uncured elastomer such as natural rubber or polychloroprene and curing the system with peroxide. Three reactions can occur: cross-linking of the elastomer, homopolymerisation of the organotin monomer, or, finally, co-vulcanisation of the elastomer and the organotin. It was found that up to 50% by weight of tributyltin acrylate could be incorporated in the rubber without any retardation in cure rate or deterioration in mechanical properties. These systems are really examples of the type of copolymer system to be described in 5.4.3.

Phillip [13, 38] has shown that the leaching rate of the toxicant depends to some extent on the nature of the elastomeric matrix. With natural rubber, T.B.T.O. is released more rapidly as its concentration in the rubber is increased, whereas tributyltin fluoride is released at a constant rate irrespective of increasing concentration. This suggests that the fluoride behaves like an inert filler. In polychloroprene coatings, both toxicants show a dependence of leaching rate on initial concentration, so that for this system the effective antifouling life cannot be extended merely by increasing the loading of organotin. Janes [39] has reported on the performance of 2.5 mm thick polychloroprene sheet containing T.B.T.O. which had been fastened to pleasure boats in 1968. The hull was virtually clean after $9\frac{1}{2}$ years, except for a few barnacles. Adhesion of the rubber to the steel hull had remained good.

5.4.3   Polymeric systems for long-term protection

The marine industry is continually seeking new methods of combating fouling and its aims are directed towards an economical coating which will provide long-term protection. Since the toxicants incorporated in the coatings are expensive, it is essential that these are utilised to maximum effect. Also, as dry-docking means loss of operating time, which in the case of giant supertankers represents a serious loss of revenue, the longer the intervals between repairs, the better. It has been calculated that extending time between dry-dockings from 24 to 30 months is worth £350,000 over a 10-year period, for a typical container vessel.

Conventional paint formulations are suitable for the majority of merchant ships which dry-dock at intervals of about 12-18 months, but these cannot offer the $2-2\frac{1}{2}$ years protection needed for tankers. For naval vessels, also, the strategic factor of constant readiness is important. Elastomeric systems whilst giving longer protection, are wasteful in that not all the impregnated toxicant is utilised and there are difficulties in attaching rubber sheets to complex profiles.

Seen in this context, the advent of polymeric antifouling systems in which organotins are attached directly to the polymer, so that they can be released slowly by hydrolysis in contact with seawater, is one of the most exciting developments in antifouling technology. Fig. 5.7 shows the mode of release of the organotin from a typical polymer.

Attack by seawater on the paint film causes hydrolysis of the organotin-ester linkage. The organotin moiety is released into the water where it can combat fouling. The depleted ester layer of the paint film, now containing hydrophilic free carboxylate groups, has little integral strength and is eroded by moving seawater, exposing a fresh surface of organotin containing polymer. This hydrolysis and erosion process is continually repeated until no paint is left on the substrate and all organotin biocide has been used up. In fact this is a simplified account of the mechanism and a great deal of development work was involved before satisfactory systems were evolved [41]. As will be seen, the key to suitable controlled erosion of the paint film was the incorporation in the system of water-soluble pigments and hydrophobic low molecular weight compounds. It was found that erosion of the paint film was then directly related to the movement of the water across it, so that an actual "self-polishing" of the surface occurred in operation. This is discussed more fully later and first some

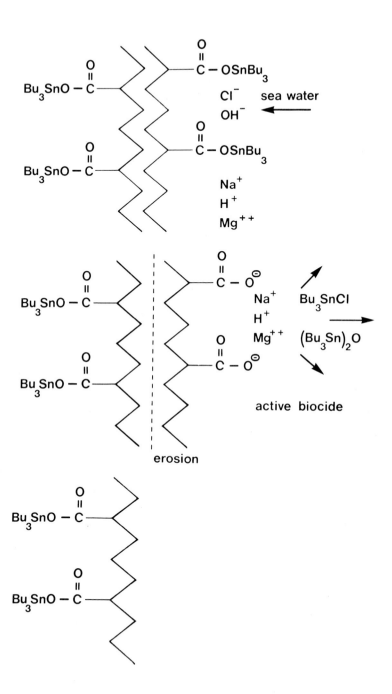

Fig. 5.7. Schematic of antifoulant release from a polymer backbone; after Gitlitz [40].

development work which has taken place with various polymers in many countries is summarised.

Polymers of trialkyltin acrylate esters were first described by Montermoso et al. in 1958 [42] in a search for thermally stable polymers. The use of organotin polymers to protect surfaces against attachment of marine organisms was patented in 1965 [43]. During the 1970s research and development in this direction were in progress in the U.S.A., Europe, Australia and Japan and by the mid-1970s, paints based on organotin polymers were available in Europe and Japan; such paints received E.P.A. approval in the U.S.A. in 1978.

Early development work by the U.S. Navy, with acrylic systems, has been described by Montemarano [44]. Acrylic polymers and copolymers are favoured for antifouling coatings because they have good film-forming properties and are compatible with other paint ingredients. The presence of acrylic groups gives the coating hydrophilic properties which improve coating performance. The most common organotin monomers are tributyltin acrylate and tributyltin methacrylate. These may be copolymerised with a number of monomers, for example alkyl acrylates and methacrylates, maleic anhydride, styrene, vinyl chloride, vinyl acetate, acrylamide and acrylic acid. One copolymer which is reported to have given excellent performance in panel tests is poly(tributyltin methacrylate/tripropyltin methacrylate/methyl methacrylate). Another approach is to esterify pendant groups in the polymer chain so as to introduce the triorganotin into the side chain. Thus, triorganotin groups are introduced into copolymers of maleic anhydride by reacting the copolymer with T.B.T.O. [45].

Subramanian and his colleagues [46] examined thermoset epoxy-based tributyltin polymers, since epoxy resins are versatile materials which can be modified to produce optimum mechanical properties. The first step was to partially esterify, using T.B.T.O., the linear polymers carrying carboxylic acid or anhydride groups. This partial ester is then cross-linked by reacting free carboxylic acid or anhydride groups with epoxy resins. A typical reaction scheme is shown in Fig. 5.8.

A variety of cross-linked polymers were prepared and tested and showed a wide range of antifouling activity in panel tests at Miami Beach, FL. Some panels remained 80% foul-free for as long as 27 months and it was considered that there was scope for achieving a high degree of optimisation with regard to mechanical properties and long-term antifouling effectiveness by suitable choice of components. Further studies concentrated on the styrene-maleic anhydride copolymer, using general principles already established. The tin content in such

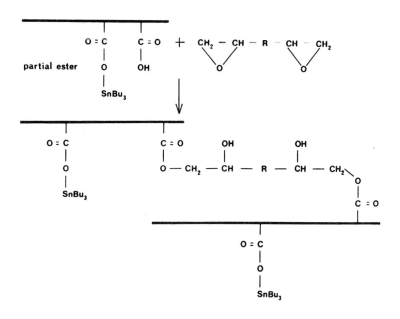

Fig. 5.8. Typical reaction scheme for producing thermoset epoxy-based tributyltin polymers [46].

polymers could be varied by controlling the degree of esterification of the base
polymer and this also controlled the degree of cross-link density obtained in
the second, curing step. Promotion of homopolymerisation at high epoxy-to-anhy-
dride ratios would have the general effect of extending the lengths of epoxy
cross-links. This could be achieved using a suitable curing catalyst. Uranyl
nitrate catalyst conferred enhanced strength and toughness on the cured polymers. Subramanian [48] also explored epoxy systems in which the tributyltin
carboxylate group was not linked directly to the backbone of the polymer chain,
but via a longer side chain. A diglycidyl ether of bisphenol-A was first modified by reacting it with tributyltin esters of ⱳ-amino acids and then curing
the resulting prepolymer with diethylene triamine at room temperature. As the
length of the side chain carrying the tributyltin moiety was increased, a reduction in strength, modulus and glass transition was observed. In addition, a
greater mobility of the tributyltin group could be expected.

Evans and Hill [49] have summarised the mechanisms of release of organotin
biocides from thermoset polymers. Diffusion of organotin from the matrix to the
surface is considered to be the rate-determining factor. Highly cross-linked
epoxy formulations decrease the diffusivity in the matrix, so that only a small

fraction of the tin would be released from these compositions during their service life. Polyurethane matrices are more permeable and are expected to release
the organotin more readily. In the case of polyesters, the concentration of the
mobile species in the matrix depends on the hydrolytic process; hence the more
hydrophilic the matrix, the greater the rate of release of the organotin. NMR
studies of the chemical nature of the tin compounds which are released from the
coatings confirms the supposition that transport of the mobile tributyltin species in the matrix is the rate determining factor.

Although much development work is being conducted with a wide range of organotin polymers, as indicated, the systems which have been most employed commercially are the acrylic polymers containing tributyltin acrylate or methacrylate.
Incorporation of additives which control the rate of breakdown of the paint
film in contact with water led to the break-through which has resulted in "self-
ablative" antifouling coatings. These developments have been reviewed [6, 41].
By controlling the polymer composition and pigmentation and by using retarders,
the rate at which the coating ablates can be regulated. Since the rate of dissolution is influenced by turbulence in the surrounding seawater and as this is
greatest at points of maximum roughness in the paint film, these peaks dissolve
more rapidly so that an overall smoothing or "polishing" of the surface takes
place. This process is confined to a reaction zone only 5 μm thick at the coating surface, but is significant enough to provide considerable savings in running costs. The lifetime of the coating is directly proportional to the thickness applied and there is continuous replacement of active surface, as shown in
Fig. 5.9.

The significance of this polishing action can be appreciated by considering
that 80% of resistance to motion of a vessel is due to frictional resistance
which depends principally on the condition of the hull and for example, 25 μm
increase in roughness can result in an extra 2.5% of power being necessary to
maintain a given speed [1]. Clearly from the nature of the polishing action,
release is greater when the vessel is moving. However, a compromise solution
between acceptable leaching during stationary periods and a practical rate of
polishing during motion has been achieved by suitable formulation. With this
type of ablative coating it is essential that manufacturers' specifications are
followed and the total protective system is built up in a series of carefully
controlled layers, the concentration of retarder in the coatings decreasing successively from first applied coat to the last, top coat.

Christie [50] has described a further development in which copolymer resins
are manufactured with a much higher weight of solids content than previously,

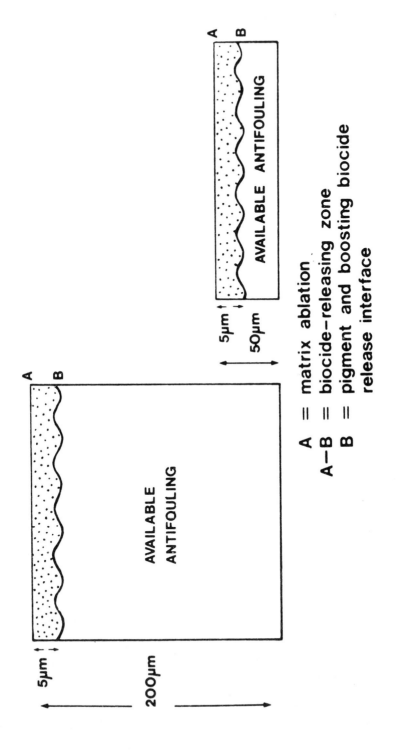

A = matrix ablation
A—B = biocide-releasing zone
B = pigment and boosting biocide
      release interface

Fig. 5.9. Self-smoothing action of "self-polishing copolymer" antifouling coatings [6].

without losing a workable viscosity. This means that fewer coatings need to be applied and good stationary performance (up to 6 months in tropical conditions) is claimed. Another advantage is that over-coatings can be applied without loss of the biocidal activity from the underlying material.

There are now a number of commercially available ablative antifouling coatings based on organotin polymers and these are employed to an ever-increasing extent on a wide range of vessels. Both the U.S. and the Australian Navies are showing interest in these systems since they reduce the dry-docking time and ensure top performance in operation.

Specific details of the proprietary formulations are not available. According to one patent [51], zinc oxide is incorporated as a seawater-reactive pigment which forms water-soluble compounds, helping to minimise localised erosion of the paint film by inducing a "planing" effect. The copolymer was specified as containing at least 50% of the organotin compound and the tin ions exchange with metal ions present in seawater to leave a water-soluble copolymer salt which then dissolves to expose fresh coating. To control the rate of ion exchange and thus to prolong the life of the coating, a hydrophobic retarder is incorporated in the paint; ideal compounds are dibutyl phthalate or dichlorodiphenyltrichloroethane, which have a water solubility of less than 5 ppm. Milne [41] has described how cuprous oxide may be used in place of the zinc oxide pigment to give the same effect, but with an added synergistic action. The pigmentation greatly increases the area of polymer in contact with water, so that lysis of the organotin occurs more readily. Part of the value of the soluble pigment might be the fact that it provides "space" for the hydrolysing polymer, since considerable conformational and volume changes can occur when the polymer changes to a salt or acid form after hydrolysis. The presence of poly-cations in seawater may allow inter- and intra-molecular bridges and serve to maintain the integrity of the diffusion layer until all or most of the organotin has hydrolysed to a given degree in the film. The residual polymer then ceases to be a film former and the leached and lysed residues of the film simply disintegrate. Calculations suggest that a five-year antifouling activity is a practical proposition with ablative coatings.

Gitlitz [40] has discussed the question of formulating organotin polymer paints, pointing out that the principles governing their performance differ from those for conventional additives. Since the coating itself is consumed when the toxicant is released, the service life depends on the erosion rate and film thickness. Major factors influencing the erosion rate of tributyltin methacrylate copolymers include:

Composition and structure of the copolymer

Levels of water-sensitive pigments

Levels of inert pigments

Level and nature of auxiliary toxicant

Level and nature of other paint additives

Pigment volume concentration

Table 5.6 gives results for film erosion obtained using a rotor apparatus to simulate seawater passing the coated surface, in the case of various copolymers [8].

Fig. 5.10. The hull of this vessel, in dry dock for maintenance painting after 24 months at sea, is clear from fouling, due to its protective ablative anti-fouling coating.
Photograph courtesy: Transocean Marine Paint Assn., Rotterdam, The Netherlands.

Incorporation of hydrophobic additives of low seawater solubility and vapour pressure have been used to control erosion rates and patents [52, 53] have claimed the use of chlorinated rubber and polyacrylate esters as erosion retarders.

Since very high levels of organotin incorporated in the coatings can affect the erosion rate and the film quality, many antifouling paint manufacturers have found it beneficial to incorporate a co-toxicant in the system, such as tributyltin fluoride or copper oxide. Doi [54] has described the development of an

TABLE 5.6

Rate of hydrolysis and film erosion in seawater rotor apparatus [8]

Composition of copolymer * %				Loss of film thickness, μm			
				2½ months		5 months	
MMA	BA	Sty.	TBMA	Clear	ZnO pigmented	Clear	ZnO pigmented
30	–	–	70	31	37	54	90
25	10	–	65	13	11	14	23
–	–	50	50	3	2	4	7

*
MMA	Methyl methacrylate
TBMA	Tributyltin methacrylate
BA	Butylacrylate
Sty.	Styrene

ablative coating in which thiocarbamates are incorporated for increased protection against fouling slimes; a slow erosion rate (5-8 μm/month) is claimed.

Finally, when discussing ablative coatings, it should be noted that when the initial hull roughness is greater than 200 μm, most companies do not recommend use of this type of antifouling, since this is the limit of the operating film thickness for many self-smoothing systems.

5.5   TESTING OF ANTIFOULING SYSTEMS

The evaluation of potential antifouling agents and of coating systems containing them, involves a whole series of test procedures ranging from simple laboratory testing to immersion trials and gathering of full-scale operating data.   Care is necessary in evaluating data to ensure that all possible parameters have been taken into consideration and that test conditions are representative of actual operating situations.

5.5.1   Laboratory studies

Christie [9] has used laboratory techniques to study spore settling and the effects of toxicants in the case of Enteromorpha. Fronds of Enteromorpha sp. were collected from shores in Devon and fruiting fronds were washed in fresh water, placed individually in specimen tubes and flooded with filtered seawater. Spores are generally released within a few minutes and samples removed for identification. Settlement of zoospores was followed by electron microscopy. To determine the effect of toxicants, aliquots of a freshly obtained suspension of zoospores were rapidly inoculated into a series of petri dishes containing 7 ml of seawater. After 5 minutes, the dishes were drained and filled with filtered

seawater. At 0.5, 5, 10, 15, 30, 60 and 120 minutes, the seawater was discarded
and replaced by triphenyltin chloride at concentrations of 0.01-0.0005 ppm in
seawater supplemented by soil extract, nitrate and phosphate. After 5 days il-
lumination sporeling counts were taken and the LD_{50} values calculated at the
different settlement times. A progressive increase in LD_{50} values with settlement time was noted. Skinner [55] developed optimum conditions for culturing
Enteromorpha sp. and Ectocarpus sp.; the toxicants under test were dissolved in
a suitable solvent and added to a known volume of seawater containing 1 ml of
spore inoculum.

More recently, Hughes [56] has developed laboratory techniques suitable for
use in the rapid screening of antifouling toxicants. He demonstrated a dose-
related alteration in the size distribution when compared to a control, of algal
suspensions of P. tricornutum and C. coccoides after 24 hours exposure to various levels of mercuric or cupric chlorides. The algal size distributions were
measured using a Coulter counter with a Channelyzer C-1000 accessory for added
ease of sampling and data analysis. In a subsequent paper Hughes [57] described
the determination of relative levels of adenosine triphosphate (ATP) from algal
samples, using the Luciferin-Luciferase assay technique. Cells exposed to toxic
levels of biocide showed a dose-related decrease of cellular ATP levels (compared to controls) after a short (24-hour) exposure period.

The rate at which toxicant is released from an antifouling coating is very
important in determining the effectiveness of that coating, and the minimum value of this leaching rate which will prevent settlement is known as the "critical
leaching rate". Determination of this rate with any precision is difficult, because of the complex and time-consuming techniques needed to analyse materials
whose leaching rates are only of the order of 1-20 $\mu g/cm^2/day$. Chromy and Uhacz
[58] have determined critical leaching rates for organotins from vinyl paints
pigmented with red iron oxide and containing 15 wt.% toxicant. Paints were applied to glass panels which were immersed in 500 ml leaching solutions (water +
2% sodium chloride solution). Leaching took place under static conditions over
150 days; the solution was changed every 10 days and the concentration of organotin in the solutions was determined. The leaching rate was plotted and after
an initial rapid leaching rate, the rate became steady. The antifouling properties of the coatings were determined in immersion tests in marine environments
and the results correlated with the leaching characteristics. The critical
leaching rate for T.B.T.O. was found to be not greater than 0.4 $\mu m/cm^2/day$ and
for tributyltin acetate was not greater than 0.7 $\mu g/cm^2/day$ under the test conditions.

It was soon realised that for more exact simulation of operating conditions some degree of relative motion between seawater and the test surface was required. A rotor test apparatus suitable for antifouling test studies has been described in a patent [51]. A mild steel disc coated overall with a conventional anti-corrosive paint and over-coated in radial stripes with the antifouling paints under test, is rotated at a peripheral speed of 38 knots in flowing seawater. De La Court [25] used a rotor apparatus consisting of a cylindrical drum upon which test panels were fixed, rotating in natural seawater at controlled pH and temperature, for his leaching tests. Paints were made up containing varying amounts of organotin compounds as well as a range of concentrations of binder, rosin and zinc oxide. After rotor ageing the paint films were analysed for tin, fluoride and zinc distribution from surface to substrate and for total residual tin content in the paint film. Again, leaching data were correlated with the results of raft tests.

Rivett [59] described an interesting approach whereby a biological method was used to detect very low leaching rates. This involved an indicator organism which was used as an indirect parameter of biocide concentration. The organism was a species of the marine, unicellular green alga Chlamydomonas sp. and the critical toxicity (minimum concentration of toxicant to inhibit growth completely) was determined for a particular toxicant in 50 ml unstirred culture medium inoculated with a known concentration of algae. The concentration of toxicant in the leachate from a paint coating was then determined by diluting the leachate until it reached the critical toxicity value under identical conditions.

Subramanian et al. [60] described a method of evaluating insoluble polymer formulations. Small holes are made in an agar medium in a petri dish and filled with the powdered polymer. The petri dishes were then incubated for a period of time, developed and the inhibition zone measured for three micro-organisms, Pseudomonas nigrifaciens, Glomerella cingulata and Sarcina lutea.

Instrumental techniques have been used to characterise paint films. Thus infra red and NMR spectroscopy have been used to study the chemical and structural properties of organotin toxicants in relation to the coating matrix, curing process, actual ageing in use and coating efficiency [61]. O'Brien et al. [62] used Mössbauer spectroscopy to identify and determine the structure of tin species in various coatings. Bird [63] described a method for determining the distribution of a biocide within a coating using a scanning electron microscope with X-ray analysis facilities. He showed how the technique could be adapted to routine examinations of paint films with computer processing of data [64]. Miale et al. [65] used neutron activation and X-ray fluorescence spectroscopy to quantify residual tin in paint films after standard exposure to seawater. Baker et al. [66] used scanning electron microscopy to examine marine coatings after underwater exposure to fouling for 6 months. The results were compared with those for laboratory salt water immersion tests. Bishop et al. [67] used reflection scanning electron microscopy to characterise microfouling occurring on vessels. Eight species of diatoms could be distinguished by this means. Kronstein [27] used infra red to characterise ether extracts from water in which antifouling coatings had been immersed. Metal content in the water surrounding the immersed coatings was also determined directly by atomic absorption spectrophotometry. The results were used to understand reactions occurring between the toxicant and the other constituents of the paint.

5.5.2   Immersion tests

Much of the preliminary screening of coatings for their antifouling activity is of an empirical nature and takes the form of immersing test panels in waters known to contain fouling species alongside untreated control panels (see Fig. 5.11). The panels are suspended from rafts at defined depths and data are normally recorded for environmental characteristics such as tidal and climatic conditions, seawater composition and temperature. Panels may be metallic or plastic and typical panel preparation comprises application of a first protective paint coating to obviate rust and corrosion problems, followed by one or more coats of the antifouling system under test. Panels are examined visually at periodic intervals and the type and extent of fouling is assessed; a photographic record is usually kept.

Beiter [29] has described tests in which immersed panels were periodically removed and small discs cut out for determination of the tin left in the coating. Antifouling efficiency could then be correlated with residual tin levels in the coating. Subramanian and Garg [46] have described the testing of polymeric antifouling systems. Poly(methyl methacrylate) plates were used as substrates and a silicone rubber frame was placed around the plate. A prepolymer/epoxide monomer reaction mixture was then poured into the cavity, covered with a glass plate, and the whole cured in an oven. The panels thus prepared were immersion tested in the normal manner. Lovegrove [68] developed a faceted raft of octagonal cross section for testing of panels. This provided simultaneously, a series of different environmental conditions for testing paint panels. The well-lit waterline would assess plant fouling; the sparsely illuminated flats were the area of settlement for barnacles and other predominantly animal fouling; and the area centred on the turn-of-bilge where light was only partially restricted, could be attacked by both plant and animal fouling. This "Turtle" raft was used to screen the activity of new antifoulant formulations in a stand-

Fig. 5.11. Test panels are used to evaluate experimental antifouling systems. Photograph courtesy: M and T Chemicals b.v., Vlissingen-Oost, The Netherlands.

ard 13-week period. Materials showing promise were then tested in the laboratory against algal spores and barnacle larvae, depending on whether their bioactivity was general or specific.

Overmars et al. [69] studied the activity of polymer-bound organotins against both algal and barnacle attack. They submerged panels at 30 cm depth, aligned horizontally, so that only algae would settle, and other panels at about 1.5 m depth in a vertical position, to assess settling of barnacle larvae. Fouling was expressed as a percentage of surface covered. All of these immersion tests were static and it was becoming increasingly realised that dynamic tests were needed to simulate actual conditions, particularly in view of the trend, for bigger vessels, towards rapid turn-around so that stationary periods are shorter. Dynamic tests were conducted by Overmars in which panels were attached to a rotor apparatus, rotating at a speed equivalent to 10 mph in continually refreshed seawater for 5 months. These tests were used to assess the erosion rate of polymer coatings and the leaching rate of toxicants. Erosion was assessed by the decrease in thickness of the coating as measured by a Permascope; leaching was determined by analysis of the surrounding water for tin moieties.

Another method of testing is patch testing, whereby a candidate coating is applied to selected locations on an actual vessel and inspected after particular time intervals (see Fig. 5.12).

Fig. 5.12. Weed control by "Intersmooth" ablative coating; weedy patch can be seen on unprotected area on the left of the control area and on numerals. Photograph courtesy: International Paint Marine Coatings, Gateshead, U.K.

These tests have the advantage of demonstrating performance under actual service conditions but are not always practicable. Coatings at least 100 m^2 in area are usually needed and the procedure is expensive, infrequent and difficult to control.

Pettis and Wake [70] have described the monitoring of a small vessel on which the entire hull was painted with a test coating. The attachment of test panels to bilge keels offers another way to screen large numbers of experimental formulations simultaneously under a reasonable approximation to service conditions, allowing the removal of replicate panels in series and their return to the laboratory for subsequent examination [41].

5.5.3   Monitoring actual performance

Most major paint manufacturers keep track of the condition of large vessels coated with their proprietary antifoulings, both for use in publicity brochures as statistical evidence of effectiveness and as an assessment of their long term practical performance for use in future system design. One major manufacturer of marine coatings, International Paint, has computerised the collection of this type of data in a system known as "International Dataplan". The operation of this system has been described by Christie [50]. International Dataplan commenced in 1977 and by April, 1983, the records of 7365 individual ships over 7500 dwt, coated by the company, had been put on file. The records are based on

detailed surveys conducted by inspectors of both new buildings and maintenance
dry-dockings. To standardise the collection of data, inspectors are issued with
a manual containing a series of technical standard diagrams which provide a vis-
ual representation of a percentage defect, either localised or scattered, covering the range from 0.1 to 100% and colour photographs of typical defects and
fouling types. Reports from the inspectors are converted into numerical data
using a fouling rating scale which takes into account the relative difficulty of
controlling various types of fouling (slime > weed > animal). Once these data
have been fed into a computer it is possible to retrieve a fouling rating associated with the hull condition of any particular vessel at dry-docking.

5.6 ENVIRONMENTAL ASPECTS OF ANTIFOULING SYSTEMS

When examining the environmental aspects of using organotin toxicants in
antifouling coatings, one has to consider several factors. Firstly there is the
question of operator safety and possible pollution during application of the
coatings to the vessels and their subsequent removal during cleaning and maintenance. Secondly there is the effect that toxicants released from the hull of
a vessel in service might have on aquatic life in general. Thirdly there is the
possible effect of environmental build-up of toxic residues.

5.6.1 Handling of organotin-based antifouling systems

The compounds which are principally used in antifouling coatings have a long
record of safety in other applications. Thus T.B.T.O. has been used for over 25
years in wood preservation without any serious cases of poisoning having been
reported. Triphenyltin compounds have had an excellent record of human and environmental safety as agricultural fungicides in the U.S.A., Europe and Asia for
over 20 years. It is thus not surprising that no cases of acute or chronic
poisoning have been reported during many years handling of organotin-based antifouling paints. Sheldon [71] has examined the effects of organotin antifouling
coatings on both man and his environment. He indicated that the main hazards in
handling relate to potential contact with eyes and with the skin and he outlined
general safety measures to be adopted when handling triorganotin antifoulants.
He stressed that chemicals such as T.B.T.O. and triphenyltin fluoride are being
safely used in shipyards around the world and their safe handling has been demonstrated. Airless spraying is recommended with full provision for protection
of the operator and nearby workmen from spray. Similar precautions should be
taken in sand blasting, welding or burning operations on ships' hulls coated
with antifouling paints.

Morgenstern [72], reporting on U.S. Navy formulations, has stated that vinyl-
based organotin antifouling paints have been used with workable safety precau-

tions by painters. Aspects of applying antifouling coatings were discussed by Banfield [73], who pointed out that the introduction of airless spraying in the 1960s for ship painting allowed high-build coatings to be applied rapidly but necessitated safeguards against spray inhalation particularly of very small diameter (5 μm) droplets which could penetrate the filters of some simple face masks. With regard to potential hazards in removing old antifouling coatings, Yoshida [74] has reported on a commercial coal-tar epoxy paint containing cuprous oxide and organotins as biocides, which, it is reported, can be freshly coated without the need for removing the old coating. It is stated to be suitable for application to ships' bottoms and to the boot topping. The ability to apply new coatings on top of old ones without losing the underlying toxicant's biocidal action has been claimed for some of the newer ablative polymeric coatings [50].

Precautions to be adopted in handling products containing organotins are specified by manufacturers, e.g. [75, 76]. They include protective clothing such as goggles, gloves and industrial footwear. Effective ventilation must be installed and with solids such as tributyltin fluoride this should prevent the fine powder from being dispersed during handling and during operations such as milling into paint formulations. Tributyltin fluoride is now often produced in pre-milled paste form for more convenient handling [40].

5.6.2   Effects of leached organotins on aquatic life

With regard to the effects of organotin toxicants leached from coatings on non-fouling species of marine life, it is known that the concentration gradient of the toxicant falls off rapidly from a lethal concentration in a very narrow layer at the coating surface, to an extremely low concentration beyond that. The low seawater solubility of most organotins and the development of coating systems allowing extremely low release rates, are other factors in favour of the organotins.

However, a number of studies have been conducted to determine the impact on marine life of organotins leached from antifouling coatings. In one study [77], invertebrates (filter feeders) and vertebrates (fish) were maintained in water containing known concentrations of tributyltin acetate. Two habitat substrates were used, one being a silt-clay mixture of high organic content and the other being largely sand with little organic matter. All vertebrates died within 3 hours at organotin concentrations of 0.15 ppm and above, and in 24 hours at 0.10 ppm. Invertebrates died between 5 and 8 days at concentrations of 0.15 ppm organotin. Clay bottoms effectively bound the organotin, rendering it unavailable to fish and invertebrates, but sand substrates were less effective. An ability of the organotin to differ in its toxicity to particular species was noted.

Linden and co-workers [78] determined the acute toxicity of 78 chemical and pesticide formulations against two brackish-water organisms, a fish (The Bleak) and a harpacticoid (Nitroca spinipes). Organotin compounds tested and their LC_{50} (I) values for 96 hours are shown in Table 5.7.

TABLE 5.7

Toxicity of organotins to aquatic organisms [78]

Organotin compound	Initial concentration killing 50% of test organisms (LC_{50} (I)) in 96 hours, mg/l	
	Bleak	Harpacticoid
Tributyltin fluoride	0.006-0.008	0.002 (0.001-0.002)
Tributyltin oxide	0.015 (0.0013-0.0017)	0.002 (0.001-0.003)
Triphenyltin fluoride	0.40 (0.34-0.46)	0.008 (0.006-0.010)

Laughlin et al. [79] have studied the action of T.B.T.O. and tributyltin fluoride on non-target organisms both as individual compounds and as leachates from two commercially available paint formulations. Marine amphipods were selected as the test organisms because of their ubiquitous occurrence and their ecological importance. The species were Orchestoides californiana from the U.S.A. and Gammarus oceanicus from Sweden. Both organotin compounds were shown to be acutely toxic at concentrations of 10 parts per billion and above. Gammarus oceanicus was exposed to tributyltin leachates from painted panels, one type of paint releasing the organotin about 10 times faster than the other. Increases in painted area and leaching rate were both directly correlated with amphipod mortality in short-term tests. Gammarus oceanicus was the more sensitive to organotins, final leachate concentrations of 4.8 parts per billion causing total mortality in 5 days. These figures are well below the value of 50 parts per billion commonly quoted for the LC_{50} value of tributyltin compounds towards aquatic organisms. The authors consider that one reason for this may be the slow-acting nature of these organotin toxicants, since toxicity only became apparent after the first 5 days of exposure in the tests with Orchestoides californiana. Intelligent choice of an antifouling paint formulation depends upon an acceptable compromise between leachates which are effective at the coating surface, but which minimise effects on non-target organisms.

There have been reports of changes occurring in the calcification process of the Pacific oyster <u>Crassostrea</u> <u>gigas</u>, introduced into France some time ago. This change manifests itself as a distortion of the shell structure. In view of the increased use of organotin-based antifouling paints on vessels harbouring where these oysters are found, a group of workers have studied the concentrations of heavy metals (zinc lead and tin) in affected molluscs, have determined the effects on calcification behaviour of leachates from painted panels containing tributyltin fluoride and have statistically correlated malformations with regions of heavy shipping [79]. They tentatively produce a mechanism by which organotins could influence calcification processes, but do not definitely establish a link between the oyster behaviour and organotin-based antifouling paints.

5.6.3 Long-term environmental effects

Although the total amount of organotin compounds released to the world's oceans by antifouling coatings is likely to be small, it can nevertheless represent a locally ecologically significant contribution. The fate of organotins in the environment is an important consideration and Chapter 10 is devoted to this subject. However, it is relevant here to summarise aspects associated with the use of antifouling paints. Recent techniques for speciation of organotin compounds in trace quantities have thrown some light on possible pathways into the environment. Many studies of the degradation of organotin compounds have demonstrated that progressive cleavage can occur in organic groups attached to tin, leading eventually to inorganic tin derivatives. Evidence is available in the literature to suggest that organotins released from antifouling systems are degraded by ultraviolet light, by soil micro-organisms and by absorption on soil. These papers have been reviewed by Blunden [80].

The possibility of biomethylation of inorganic tin residues in the environment has been considered and in certain studies [81] trimethyltin hydroxide was added to anoxic estuarine sediments; headspace samples were drawn off periodically and analysed for tetramethyltin by gas chromatography. The formation of tetramethyltin from trimethyltin hydroxide was demonstrated in anaerobic estuarine sediments but not in seawater and not as surface effects on bentonite. However, the low yield was such that these reactions were not considered to be very significant, particularly in the case of the tributyltins.

A novel approach has been described by Porter [82] who is engaged in growing strains of <u>Pseudomonas sp</u>. bacteria on agar surfaces treated with various organotin compounds, to determine whether the bacteria can transform the tin compounds. The atmosphere above the bacteria is being analysed to determine whether the bacteria are metabolising the tin and exhaling it in some other form. A number of workers have attempted to study the potential environmental impact of organotin compounds leached from antifouling paints. Good and colleagues [31] have described a programme of work which is to be conducted for chemical and physical characterisation of the coatings. Molecular spectroscopy will be used to characterise the toxicant in the bulk of the coatings, surface techniques (ESCA, AES and backscatter Mössbauer spectroscopy) will attempt to determine the organotin species at the surface, photomicroscopy will be used to monitor gross surface features as a function of coating methods, curing and ageing, and finally various analytical procedures will be adopted to determine the amount and chemical nature of, the tin species released to the environment.

The difficulty of determining organotins at the very low levels needed for meaningful studies, has hampered many researchers. Jewett [83] has reviewed new methods for speciation of bio-active organotin leachates. By using "reverse bonded phase" high performance liquid chromatography techniques, triorganotins can be collected from aquatic media in parts per billion concentrations for subsequent characterisation by graphite furnace atomic absorption spectrophotometry.

Monaghan et al. [84] studied the degradation of organotins released from antifouling paints in water by suspending triphenyltin and tributyltin compounds in distilled water and in artificial seawater and recovering dissolved components by extraction with chloroform. Residues from the extracts were examined by thin layer chromatography and infra red spectroscopy. In the case of tributyltin compounds no other moieties were identified but with the triphenyltin compounds, tetra-, tri-, di- and mono-phenyltin groups were found. Structures such as R_3SnOH and $R_3SnOSnR_3$ can be eliminated as possible soluble species since O—H and Sn—O—Sn stretching vibrations were not found in the infra red spectra of the chloroform extracts.

Another possible mode of ingress of organotins into the environment is through the disposal of organotin-containing grit which results from abrasive blasting of coated ships' hulls. Harris et al. [85] studied the feasibility of using such waste for land filling and determined the leaching behaviour and migration of organotin compounds through clay, topsoil and sand, using waste-contaminated grit buried in lysimeters. Both horizontal and vertical migration was observed for the clay and topsoil whereas vertical migration was predominant in sandy soil. All soils had a strong affinity for organotins and soil disposal of waste was considered an effective interim method of dealing with contaminated grits. Tests on clay-based soils conducted by Slesinger [86] indicated that 95-99% of applied organotin compound remained where it was placed, suggesting that leaching from land fill soil is highly improbable and that this method of disposal would pose no threat to the environment. Degradation in soils has been studied using carbon-14 labelled triorganotin compounds. In one study [87] 50% of applied T.B.T.O. disappeared from silt and sandy loams in about 15 and 20 weeks, respectively; dibutyltin derivatives and carbon dioxide appeared to be degradation products. In an earlier study, Barnes et al. [88] estimated the half-life of triphenyltin acetate in soil to be 140 days.

A patent [89] has described a method of detoxifying organotin-containing paint residues by treating them with nitric acid. Laboratory experiments showed a complete conversion of the organotin to non-toxic inorganic tin. This was confirmed in toxicity studies using fish, in water which was contacted with treated waste. Conversion was complete, using 20% acid, in 3 days at 35°C and in one week at 20°C.

5.7 COMMERCIAL DEVELOPMENTS

The preceding sections have all examined technical aspects of organotin-based antifouling coatings, but to put these into perspective, it is useful to have an over-view of the commercial developments in these systems and of the economic and marketing incentives that have led to their extensive adoption by the world's shipping.

Before 1960, the basic type of antifouling paint consisted of copper salts, usually cuprous oxide, added to vinyl and other paint resins. These paints had been marketed by every major paint company for the marine industry and sold in retail stores for use on pleasure boats. The U.S. Navy's own formulations were also based on this system. These paints were, however, of comparatively short life and because of the high leaching rate, considerable quantities of copper salts were released to the ocean. Another, practical problem was in getting sufficient binding of the paint film with high loadings of cuprous oxide.

In the early 1960s when the powerful biocidal action of the trialkyltin and triaryltin compounds became known, T.B.T.O. was introduced into antifouling paints, both as a booster for cuprous oxide and in its own right, in vinyl and acrylic paint systems. Brightly coloured paints could be obtained, which were particularly attractive for the pleasure boat industry. Moreover such paints could be used on aluminium-hulled boats without the risk of bimetallic corrosion occurring between aluminium and copper. Although T.B.T.O. gave adequate protection for small boats which were laid up at the end of a season, the leaching rate was too high for the longer term protection required in commercial vessels. The introduction of solid organotins such as tributyltin and triphenyltin fluorides, either alone or in combination with T.B.T.O., enabled longer antifouling protection to be achieved in service. A variety of effective commercial products was developed and control of shell and algal settlement for 18-24 months was achieved on fast, quick-turnaround vessels.

During the 1960s there was a dramatic change in shipbuilding methods and in the size of vessel. This increased the demands put on the antifouling coating. Protection against algal fouling such as the green sea grass Enteromorpha intestinalis and the brown grass Ectocarpus sp. became a prime necessity for large tankers whose huge bottoms provided scope for heavy fouling and consequent increased fuel costs. Specific toxicants active against this type of organism were incorporated in antifouling paints to combat algal fouling.

In 1974 the first ablative polymer coating was introduced, representing a significant development in antifouling technology. Whereas with conventional paint systems the service life is proportional to the logarithm of the film thickness, with these polymer coatings, service life is a direct linear function of film thickness. Moreover, since erosion of the paint film is controlled by contact with moving seawater, a self-polishing action is obtained. The advantages of ablative systems have been summarised as follows:

Lifetime proportional to thickness applied
Overcoating without loss of existing activity
Avoidance of coatings build-up on hulls
Efficient utilisation of toxicant in paint
Continuous provision of an active surface
Automatic self-polishing action

The first system to be developed was the SPC (Self Polishing Copolymer) system marketed by International Paint Marine Coatings. It received full E.P.A. registration in the U.S.A. in 1978 and was immediately accepted by the shipping industry there. The Queen Elizabeth II was first painted with the SPC system in 1978, providing estimated fuel savings of 12%. Further development work has led to the next generation of ablative coatings, the "Intersmooth" series produced by International Paint. Progress with these coatings has been reviewed [41, 50].

Ablative coatings are now being produced commercially by many paint manufacturers all over the world; a list has been published [90]. Products include those shown in Table 5.8.

To indicate the scale of acceptance of these coatings, one manufacturer

Fig. 5.13. An ablative antifouling coating being applied to the hull of a large vessel.
Photograph courtesy: Hempels Marine Coating A/S, Lyngby, Denmark.

claimed that by the end of 1982, some 164 sea-going trading vessels, representing 9 m tonnes of shipping, were using its ablative coatings. Another supplier has estimated cost savings with its antifouling system at around £87,000 per 250 days' service per annum; investment in the coating could be returned by fuel savings within 6 months in service at current dock and material prices. One vessel protected with self-polishing coatings is reported to have actually shown a slight increase in operating speed after 12 months service [41].

The speed of developments in organotin-based antifouling paints over the last few years suggests that the definitive coating has still to be attained. However the degree of protection which can now be afforded to shipping represents a significant increase in operating efficiency.

TABLE 5.8

Commercially available ablative coatings [90]

Trade name	Company
Torpedo SPC Antifouling	Berger Torpedo Marine Paints, London, U.K.
Camrex C-Pol Camrex C-Speed	Camrex, Sunderland, U.K.
Intersmooth SPC	International Paint plc, Gateshead, U.K.
Sigmaplane Antifouling	Sigma Coatings, London, U.K.
Nautic Modules	Hempels Marine Paints A/S, Lyngby, Denmark.
Devoe ABC AF System	Devoe Marine Coatings Europe, Rotterdam, The Netherlands.
Drag Reducing Polymer	Transocean Marine Paint, Rotterdam, The Netherlands.
Takata LLL Antifouling	Jotun Marine Coatings, Sandefjord, Norway.
Starwide SP 2000	Star Maling OG Lakkfabrikk, Drammen, Norway.
Takata Sea Queen	Nippon Oil and Fats Co. Ltd., Tokyo, Japan.
SRA Series	Korea Chemical Co., Seoul, Korea.

REFERENCES

1   F. Dawans, Rev. de L'Inst. Français Petrole, 37 (1982) 767-807.
2   J. Engelhart, C. Beiter and A. Freiman, Proc. 1st Int. Contr. Rel. Pestic. Symp., Akron, OH, U.S.A., (1974) 17 pp.
3   W.A. Corpe, Proc. 4th Int. Congr. Marine Corr. Foul., France, (1976) 105.
4   G.A. Janes, Proc. 2nd Int. Contr. Rel. Pestic. Symp., Dayton, OH, U.S.A., (1975) 293.
5   G.J. Biddle, Paper pres. at 3rd R.A.N. Underwater Paint Symp., Melbourne, Australia, (1973).
6   A.O. Christie, J. Oil Col. Chem. Assn., 60 (1977) 348-353.
7   L.V. Evans, Botanica Marina, 24 (1981) 167-171.
8   D. Atherton, J. Berborgt and M.A.M. Winkler, J. Paint Technol., 51 (1979) 88-91.
9   A.O. Christie, Proc. 3rd Int. Congr. Marine Corr. Foul., Gaithersburg, MD, U.S.A., (1972) 674-678.
10 Catalogue of Main Marine Fouling Organisms, O.E.C.D. Pub., Vol. 1: Barnacles; Vol. 2: Polyzoa; Vol. 3: Serpolids; Vol. 4: Ascidians; Vol. 5: Sponges; Vol. 6: Algae.
11 W.H. Tisdale, Brit. Pat. 578,312 (1943).

12 R.F. Bennett and R.J. Zedler, J. Oil Col. Chem. Assn., 49 (1966) 928-953.

13 A.T. Phillip, Austral. Oil Col. Chem. Assn. Proc. and News, July (1973) 17-22.

14 J.M. Barnes and L. Magos, Organomet. Chem. Rev. 3 (1968) 137-150.

15 W.N. Aldridge, in J.J. Zuckerman (Ed.), Organotin Compounds, New Chemistry and Applications, A.C.S., Adv. Chem. Ser. 157 (1976) 186-196.

16 M.J. Selwyn, ibid., 204-226.

17 W.T. Hall and J.J. Zuckerman, Inorg. Chem., 16 (1977) 1239-1241.

18 N.F. Cardarelli, Controlled Release Molluscicides, Environ. Manag. Lab. Mon., Univ. Akron, OH, U.S.A., (1977) 136 pp.

19 P.A. Millner and L.V. Evans, Plant, Cell and Environ., 3 (1980) 339-347.

20 A.P. Boulton, A.K. Huggins and K.A. Munday, Toxicol. Appl. Pharmacol., 20 (1971) 487-501.

21 Albright and Wilson Ltd., Tech. Service Note 1976 Feb. 10, 9 pp.

22 R.D. Mearns, J. Oil Col. Chem. Assn., 56 (1973) 353-362.

23 N.A. Ghanem and M.M. Abd-El-Malek, Corr. Contr. Coat. Meetg., Bethlehem, PA, U.S.A., (1978) 399-410.

24 N.A. Ghanem and M.M. Abd-El-Malek, J. Coat. Tech., 5 (1979) 29-35.

25 F.H. De La Court and H.J. De Vries, Proc. 4th Int. Congr. Mar. Corr. Foul., France, (1976) 113-118.

26 J. Lorenz, J. Oil Col. Chem. Assn., 56 (1973) 369-372.

27 M. Kronstein, Ind. Eng. Chem. Prod. Res. Dev., 20 (1981) 5-12.

28 H. Plum, F. Runggas and M. von Haaren, Ger. Pat. 2,240,487 (1974).

29 C.B. Beiter, Proc. Int. Contr. Rel. Pestic. Symp., Akron, OH, U.S.A., (1976) II-22 - II-47.

30 C.P. Monaghan, J.F. Hoffman, E.J. O'Brien, L.M. Frenzel and M.L. Good, Proc. Int. Contr. Rel. Pestic. Symp., Corvallis, OR, U.S.A., (1977) 1-7.

31 M.L. Good, C.P. Monaghan, V.H. Kulkarni and J.F. Hoffman, A.C.S. Div. Org. Coat. Plast. Chem., 39 (1978) 578-581.

32 N.F. Cardarelli, Controlled Release Pesticides Formulations Publ. C.R.C. Press Inc., Cleveland, OH, U.S.A., 1976 210 pp.

33 S.V. Kanakkanatt and N.F. Cardarelli, Int. Congr. Pure Appl. Chem., Macromol. Reprint, 1 (1971) 614.

34 N.F. Cardarelli, Tin and its Uses, 93 (1972) 16-18.

35 J.M.D. Woodford, Austral. Def. Sci. Serv., Def. Std. Labs. Report No. 496, 1972 13 pp.

36 A. De Forest, R.W. Pettis and A.T. Phillip, Austral. Def. Sci. Serv. Def. Std. Labs. Report No. 589, 1974.

37 P. Dunn and D. Oldfield, Paper to 3rd Austral. Rubber Technol. Conven., Inst. Rubber Ind., Australas. Sect., Terrigal, N.S.W., Sept. 18-21st, 1974 30 pp.

38 A.T. Phillip, Progr. Org. Coatings, 2 (1973/1974) 159-192.

39 G.A. Janes and R.L. Senderling, Proc. 6th Int. Symp. Contr. Rel. Bioact. Mat., New Orleans, LA, U.S.A., 1979 II-3-II-6.

40 M.H. Gitlitz, J. Coatings Technol., 53 (1981) 46-52.

41 A. Milne, Royal Inst. Chem. Ann. Chem. Congr., Durham, April, 1980, 23 pp.

42 J.C. Montermoso, T.M. Andrews and L.P. Marinelli, J. Polymer Sci., 32 (1958) 523-525.

43 J. Leebrick, U.S. Patent 3,167,473 (1965).

44 J.A. Montemarano and E.J. Dyckman, J. Paint Techn., 47 (1975) 59-61.

45 J. Kowalski, W. Stanczyk and R. Chojnowski, Rev. Si, Ge, Sn and Pb Compds., 6 (1982) 225-286.

46 R.V. Subramanian and B.K. Garg, Proc. Int. Contr. Rel. Symp., Corvallis, OR, U.S.A., (1977) 154-164.

47 R.V. Subramanian and M. Anand, in S.S. Labana (Ed.), Chemistry and Properties of Crosslinked Polymers, Publ. by Academic Press, NY, U.S.A., (1977) 19 pp.

48 R.V. Subramanian, R.S. Williams and K.N. Somasekharan, A.C.S. Div. Org. Coat. Plast. Chem., 41 (1979) 38-43.

49 C.J. Evans and R. Hill, Rev. Si, Ge, Sn and Pb Compds., 7 (1983) 57-125.

50 A.O. Christie, Paper to Ship Repair and Maintenance 83 Confce., New Orleans, U.S.A., June 1st-3rd, 1983, 7 pp.
51 A. Milne and G. Hails, Brit. Pat. 1,457,590 (1976).
52 Ger. Pat. 2,843,955 (1979).
53 Netherlands Pat. 78-10141 (1979).
54 H. Doi, Zosen, 26 (1982) 34-40.
55 C. E. Skinner in A.H. Walters and E.H. Heuck-van-der-Plas (Eds.), Biodeterioration of Materials, Vol. II, Appl. Sci. Publ. Ltd., London, (1972) 456.
56 B.C. Hughes, Marine Algae in Pharmaceutical Science, 10th Int. Seaweed Symp., Publ. Walter de Gruyter and Co., Berlin - New York, In Press.
57 B.C. Hughes, Proc. 5th Int. Biodet. Symp., Aberdeen, (1981) In Press.
58 L. Chromy and K. Uhacz, J. Oil Col. Chem. Assn., 61 (1978) 39-42.
59 P. Rivett, J. Appl. Chem., 15 (1965) 469-473.
60 R.V. Subramanian, B.K. Garg and J. Corredor, in C.E. Carraher, J.E. Sheats and C.V. Pittman (Eds.), Organometallic Polymers, Publ. Acad. Press, London, (1978) 181-194.
61 J.F. Hoffman, K.C. Kappel, L.M. Frenzel and M.L. Good, ibid., 195-205.
62 E.J. O'Brien, C.P. Monaghan and M.L. Good, ibid., 207-218.
63 R.J. Bird, J. Oil Col. Chem. Assn., 60 (1977) 256.
64 R.J. Bird and D. Park, J. Oil Col. Chem. Assn., 61 (1978) 151-156.
65 J.B. Miale, G. Gleason, R.P. Porter, D.H. Hesse and T.A. Mende, Proc. 5th Int. Symp. Contr. Rel. Bioactive Mat., Gaithersburg, MD, U.S.A., (1978) 12 pp.
66 F.L. Baker, L.H. Princen and M. Kronstein, ibid., 11 pp.
67 J.H. Bishop, F. Marson and S.R. Silva, Dept. Supply, Austr. Def. Sci. Serv., Def. Stand. Labs., Tech. Note 224 (1972) 6 pp.
68 T. Lovegrove in D.W. Lovelock and R. Davies (Eds.), Techniques for the Study of Mixed Populations, Soc. Appl. Bacteriol. Tech. Ser., 11 (1978) 63-69.
69 H.G.J. Overmars, F.H. De La Court and J.F.A. Hazenberg, Proc. 5th Int. Congr. Mar. Corr. Foul., Spain, (1980) 99-112.
70 R.W. Pettis and L.V. Wake, U.S. Dept. Comm., Nat. Tech. Inf. Serv., Tech. Note 377 (1975) 10 pp.
71 A.W. Sheldon, J. Paint Technol., 47 (1975) 54-58.
72 E.A. Morgenstern, Paper presented at 18th Ann. Mar. Coat. Conf., Monterrey, CA, U.S.A., (1978) 11 pp.
73 T.A. Banfield, Ind. Fin. Surf. Coat., 22 (1970) 4, 5, 7, 15.
74 M. Yoshida, Ship. Mar. Eng. Int., 103 (1980) 258-260.
75 Anon., Albright and Wilson, Tech. Note, (1976).
76 Anon., Advice on Toxicology and Sage Handling of Organotin Compounds, Schering AG Publication, 20 pp.
77 M.L. Good, D.S. Dundee and G.S. Swindler, Proc. 6th Int. Symp. Contr. Rel. Bioactive Mat., New Orleans, LA, U.S.A., (1979) II-1-II-2.
78 E. Linden, B.E. Bergtsson, O. Svanberg and G. Sundstrom, Chemosphere, 8 (1979) 843-851.
79 C. Alzieu, M. Héral, Y. Thibaud, M.J. Dardignac and M. Feuillet, Rev. Trav. Inst. Pêches Marit., 45 (1982) 1-26.
80 S.J. Blunden, L.A. Hobbs and P.J. Smith, SPR 35 (Environmental Chem.) Volume 3, Chapter 2 Environmental Chemistry of Organotin Compounds, 1984, 49-77.
81 H.E. Guard, A.B. Cobet and W.M. Coleman, Science, 213 (197() 770-771.
82 G. Porter, Dim./NBS, 1980, Mar., 6 pp.
83 K.L. Jewett, W.R. Blair and F.E. Brinckman, Proc. 6th Int. Symp. Contr. Rel. Bioact. Mat., New Orleans, LA, U.S.A., (1979) II-13-II-17.
84 C.P. Monaghan, V.I. Kulkarni, M. Ozcan and M.L. Good, U.S. Gov. Rep., Tech. Rep. No. 2, AD-A087374 (1980) 30 pp.
85 L.R. Harris, C. Andrews, D. Burch, D. Hampton and S. Maegerlein, U.S. Dept. Navy, Ship Mat. Eng. Dept., Res. Dev. Rep., No. DTNSRDC/SME-78/2A (1979) 57 pp.
86 A.E. Slesinger, Proc. 17th Mar. Coat. Conf., Nat. Paint Coat. Assn., Mississippi, U.S.A., (1977) 16 pp.
87 D. Barug and J.W. Vonk, Pesticide Sci., 11 (1980) 77-82.

88 R.D. Barnes, A.T. Bull and R.C. Poller, Pesticide Sci., 4 (1973) 305.
89 T. Dowd, U.S. Pat. 4,216,187 (1980).
90 Anon., Lloyds Ship Manager, 3 (1982) 27-30.

Chapter 6

AGRICULTURAL CHEMICALS

6.1   INTRODUCTION

Plants are one of the three main life forms on earth, the others being animals and humans; moreover the latter two forms depend on plants for their existence, either directly as a source of nourishment or indirectly as producers of
oxygen.

In order to utilise plant forms as food resources, man began cultivating
crops and the manner in which this cultivation has developed over the years has
resulted in more efficient agricultural methods but also in some cases in increased vulnerability to disease of the crops. This is because the dedication
of large areas of land solely to one crop, essential for cost-effective land
management, has provided ideal conditions for colonies of insects and fungi to
grow rapidly and to damage or even destroy the crop. In order to prevent, or at
least restrict such harm, specific protective measures, including the use of
plant protection agents, are necessary right from the beginning of cultivation.

There has been much controversy concerning the use of pesticides which are
variously claimed to "upset the ecological balance of nature" or to "introduce a
long-term burden of toxic substances in the environment", yet clearly without
responsible control, our vital food resources would be decimated, resulting in
wholesale starvation. As a European example of what crop disease can do, there
is the great potato famine of 1840 in Ireland, in which thousands of people died
and as a result of which, millions emigrated. This was caused by potato blight,
Phytophthora infestans which today can be effectively controlled by chemical
agents, as will be described later. Even now, it has been stated [1] that about
a third of the world's food production is lost due to pests and bad weather conditions, most of this loss occurring in the Third World where food is needed
most. The simple fact should be remembered that the primary function of a pesticide is not to kill, but to protect life.

Chemical control methods have proved very effective in combating pests and
particularly those treatments based on organometallic compounds. Careful
screening of candidate compounds for toxicity, involving chemical, biological
and environmental testing, regulation of dosages and loadings on crops, together
with regular monitoring of environmental levels of the toxicant, including res-

idues in treated crops or other stages of the food chain, all ensure that pesticides are used safely and responsibly. These procedures are described in more detail in 6.2. Specifications for pesticides for agricultural use are published by the Food and Agricultural Organisation (F.A.O.) of the United Nations in the form of booklets drawn up by the F.A.O. Working Party of Experts on the Official Control of Pesticides. There are also national controls by bodies such as the Environmental Protection Agency in the U.S.A. and the U.K. Government's Health and Safety Executive which operates a Pesticides Safety Precautions Scheme.

6.1.1 Introduction of organotin compounds

Organotin compounds were first introduced into crop protection in the early 1960s and today a number of commercial formulations based on triorganotin compounds are available and are finding increasing use in a number of areas. This is because they are not only potent biocides with a specificity against the target organism, but have the advantage that they ultimately degrade to harmless inorganic forms of tin, thus leaving no toxic residues in the environment. Fig. 6.1 shows the organotin pesticides currently in use in agriculture. The characteristics of these compounds are described in more detail later.

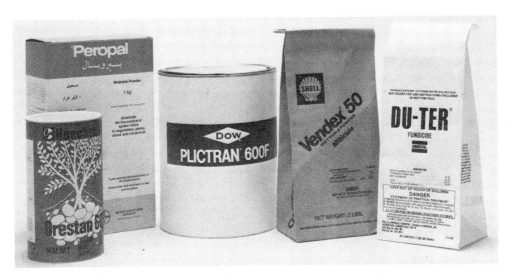

Fig. 6.1. Commercially available pesticides based on organotin compounds. See text for details.

Although the biological activity of organotin compounds had been known as early as 1929, it was work at the Institute for Organic Chemistry TNO, Utrecht which showed the powerful fungicidal properties of triorganotin compounds, which sparked off interest in their potential for crop protection. In the case of the trialkyltin compounds, their toxicity to plants was too high for practical use in agriculture but the triphenyltins showed a sufficiently wide margin between fungitoxicity and phytotoxicity to enable them to be used safely. A preparation based on triphenyltin acetate was submitted for testing in F. R. Germany against Phytophthora in potatoes in 1953-1954 and following successful performance, was authorised for use. This product, trade-named "Brestan" * was introduced by Hoechst and was the first organotin compound to be used in crop protection. Shortly afterwards a product based on triphenyltin hydroxide and trade-named "Du-Ter" * was developed and marketed by the Dutch firm Philips-Duphar. The use of these compounds was subsequently extended from protection of potatoes against blight to the control of disease on other crops; triphenyltin acetate has proved particularly effective in protecting sugar-beet. In the mid-1960s the powerful acaricidal properties of tricyclohexyltin compounds were discovered, leading to the commercial development of "Plictran" * by Dow Chemical Company. This organotin miticide was followed by others: fenbutatin oxide developed by Shell and marketed as "Vendex" * in the U.S.A. and as "Torque" * elsewhere; and 1-tricyclohexylstannyl-1,2,4 triazole developed by Bayer AG and marketed as "Peropal" *. Today, usage of these organotin pesticides has reached several thousands of tons a year worldwide and this figure is growing.

6.2 DEVELOPMENT OF AN AGRICULTURAL CHEMICAL

Each year many thousands of chemicals are synthesised and tested for use as potential pesticides. Yet many obstacles lie in the path of getting a laboratory chemical into practical use. Van der Kerk [2] estimated that in 1951 only 1 in 2000 compounds reached practical application; with increasingly stringent requirements and more sophisticated testing, this situation has worsened and Huber [3] stated that in 1977 as many as 8000 to 10,000 compounds had to be tested to arrive at one successful formulation. In general it takes 8-10 years for the first kilogramme of compound to be ready for practical application. Not only must a candidate compound have intrinsic toxicity to parasitic organisms and low toxicity towards the host plant, but it must also be stable under environmental conditions (humidity, oxygen, light) for the length of its action. It must also have low toxicity for humans and animals both during application to crops and in regard to harvested products. Economic factors are also important.

Huber [3] has described the detailed stages of development needed for achieving a successful agricultural pesticide. The first stage consists of a survey of the patent situation and a literature survey, after which compounds are syn-

---

* Trade name throughout.

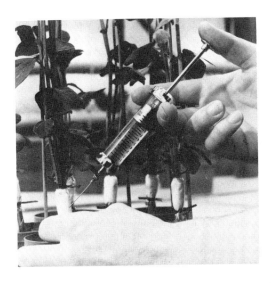

Fig. 6.2. Making localised application of a test solution to the stem surface to ascertain whether it is taken up and transmitted to the leaf.
Photograph courtesy: Hoechst AG, F. R. Germany.

thesised in the laboratory and tested against insects, fungi, algae and as herbicides. Rough preliminary screening tests, with the candidate compound sprayed as powder or in aqueous suspension, are conducted in laboratories and greenhouses. The next step for those compounds which have shown promising activity, is to determine the lowest concentration at which they are active and the approximate range of effectiveness. Discussions are held with biologists, farmers, chemists and formulation experts after which toxicity tests are set up with reference to mammals, fish and animal life eaten by fish and methods of analysis are decided on, both for the compound and for its possible breakdown products. Primary field tests are initiated to determine the activity and breadth of application under different environmental conditions, long-term behaviour against plants and possible side effects under natural weather conditions. The minimum amounts of the compound for optimum effectiveness are also established.

Simultaneously with field tests, toxicity studies are conducted. A large body of favourable data is needed in order to convince the relevant authorities of the safety of introducing a new product, so that permission can be obtained for its use. These studies include determination of acute toxicity (one application at a high level), sub-chronic (several applications at lower levels) and chronic toxicity (many applications at low levels over extended periods), against mammals (typically dogs and rats), fish and birds. The fate of the

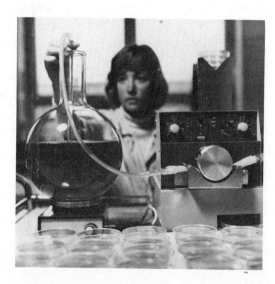

Fig. 6.3. Using sensitive equipment to determine residues on treated materials.
Photograph courtesy: Hoechst AG, F. R. Germany.

compound in the food chain is also established, for example algae may be placed
in water containing the compound; these algae are eaten by small crabs and the
latter eventually by fish. By examining the organs of the creatures it can be
established whether there is any accumulation of compound in the chain. Further
experiments determine the influence of the compound on growth and multiplication
as well as on the photosynthesis and breathing of the algae.

When all the field and laboratory tests are completed, the compound is passed
from the research department to the agricultural development department which
undertakes world-wide full-scale field tests in the principal agricultural zones
under varying climatic and soil conditions. Later, especially in F. R. Germany,
official testing authorities participate in the test programme and all data are
filed and statistically processed by computer. All these test crops must be
destroyed until clearance for use of the candidate compound is obtained. The
results are used to classify the compounds in toxicological tables, to determine
the permissible amount of residues on, and in, the crop as well as to determine
the permissible time interval between application and harvesting. Huber adds
that in F. R. Germany the Biologische Bundesanstalt (BBA) and the Bundesgesund-
heitsamt (BGA) can also request three-generation tests with rats in order to determine whether there is any evidence of malformation, hereditary effects or
cancer associated with the use of the compounds. Normally the BBA agrees to the
introduction of a new compound only after it has conducted its own tests. This

series of studies and trials accounts for the time interval of 8-10 years required for introduction of a new pesticide.

The World Health Organisation is also involved in the assessment and control of pesticides and in the drafting of international standards. The chemical and biochemical methodology adopted in assessing hazards of pesticides for man, have been summarised in a W.H.O. report [4]. The occurrence and severity of toxic effects are related to the dose. For pesticides this is greatest in the short term for formulators, mixers, and those applying the chemical to crops. Phased introduction of a pesticide, which is essential for determining its ability to control a particular crop disease vector, also provides an opportunity to seek the earliest evidence of toxic effects amongst those most heavily exposed. Since skin contamination is the most important source of exposure, the dermal toxicity of a compound is frequently the limiting factor in preventing its further development.

The hazards of any new type of insecticide can be assessed only on the basis of some knowledge of its action and metabolism in mammals. Analytical techniques are used to determine the magnitude of contamination of non-target materials, as well as for controlling the dosage on the pest itself. The amount of pesticide and its products of degradation or metabolism are assessed chemically, whilst biochemical techniques are used to evaluate effects caused by the pesticide or its metabolic products after absorption. When a compound has passed all its trial stages and is introduced into operational use, a monitoring schedule is still implemented to determine amounts being released into the environment or absorbed by workers applying the pesticide. Direct methods involve measuring inhalation or dermal exposure, but these do not measure actual absorption of the agent. In some cases, the degree of exposure of man to a pesticide may be evaluated by measuring the concentration of parent compound and of any biotransformation products in biological samples, for example in blood, urine and adipose tissue. Analysis of samples for pesticide residues can yield valuable data on the distribution, persistence and possible harmful effects of a given pest control programme on man and his environment.

In the case of the commercially adopted organotin-based pesticides, considerable data on their toxicological effects were collected before their introduction and these are summarised later.

6.3   ORGANOTIN-BASED PESTICIDES

6.3.1   Triphenyltin compounds

The trialkyltin compounds are extremely powerful biocides but unfortunately this toxicity extends also to plants. Triaryltin compounds on the contrary, whilst still being fungitoxic, are much less damaging towards many plants and triphenyltins were the first organotins to find use in crop protection. Physical properties of the triphenyltin compounds used as pesticides are shown in Table 6.1. The International Standards Organisation common name for triphenyltins is "fentin", e.g. fentin acetate.

A detailed review of triphenyltin compounds, covering their physical, chemical and biological properties, toxicology analysis and environmental behaviour, as well as their use in agriculture, has been produced by Bock [5].

TABLE 6.1

Physical properties of triphenyltin compounds [5]

Compound	Appearance	Melting point °C	Solubility characteristics
Triphenyltin acetate	White crystalline powder	119-124	Practically insoluble in water and only slightly soluble in alcohols and aromatic solvents
Triphenyltin hydroxide	White crystalline powder	118-124	Practically insoluble in water, soluble in benzene, methanol and other common organic solvents
Triphenyltin chloride	White crystalline powder	103-107	Practically insoluble in water, soluble in aromatic solvents and in chlorinated hydrocarbons

Triphenyltin chloride in the presence of water quickly hydrolyses to the hydroxide; the acetate in an aqueous suspension at 20°C hydrolyses almost completely in a few hours, the presence of acetic acid slowing down this hydrolysis. In aqueous solutions or suspensions at room temperature, triphenyltin compounds decompose slowly, losing the phenyl groups as benzene. On boiling, the triphenyltin content is diminished within two hours (depending on pH) to about 40-70% of the original concentration. Heating triphenyltin compounds with acids or alkalis leads to rapid decomposition. Depending on the anion, triphenyltin compounds vary in their thermal stability. Triphenyltin chloride remains unchanged for 10 hours at 95°C and will not decompose unless heated at above 270°C. Data on the thermal decomposition of triphenyltin acetate are inconsistent. At normal temperatures under dry conditions, the compound does not degrade, but minor decomposition has been noted after a few hours at 45°C, increasing at 100°C and this has been attributed to atmospheric moisture causing hydrolysis. Triphenyltin hydroxide loses water at fairly low temperatures but is fairly stable below 45°C. At 60-70°C it transforms rapidly to hexaphenyldistannoxane. The results of storage tests on some triphenyltin compounds in sealed bottles at 45°C are shown in Table 6.2.

TABLE 6.2

Degradation of triphenyltin compounds at 45°C [5]

Duration of test Weeks	Active ingredient %			
	Acetate	Hydroxide	Chloride	Oxide
0	98.3	98.7	98.1	99.3
1	97.8	98.1	98.0	98.4
2	97.2	97.5	98.0	97.5
3	97.0	97.3	97.9	97.4

Triphenyltin compounds have very good adhesive properties and are retained firmly by leaves on which they have been sprayed, even after heavy rainfall.

A preparation under the name "V.P. 19-40" was developed by Hoechst in F. R. Germany and received official trials in 1953-54. Following satisfactory performance, the preparation was officially tested by the BBA in Germany and acknowledged to be effective at concentrations of 1.8 kg per hectare (in a 0.3% spray) against Cercospora beticola in sugar-beets and against Phytophthora infestans in potatoes. The preparation was registered under the trade name "Brestan" and introduced for commercial use in the early 1960s. Brestan was found to be equally effective against the same range of fungi as copper-based fungicides but at about one tenth of the dosage [6].

Shortly afterwards triphenyltin hydroxide was introduced as a pesticide by Philips-Duphar in The Netherlands under the trade name "Du-Ter". Like the acetate, the compound was found to be active at much lower concentrations than were copper-based fungicides. These triphenyltin compounds operate by contact and their action is mainly protective rather than curative. They exert no systemic action and are not absorbed or translocated by plant tissue. Triphenyltin compounds have a broad spectrum of activity against pathogenic fungal organisms and also have some algicidal activity. They are being successfully used for disease control on many crops, including beans, carrots, celery, cacao, coffee, ground nuts, hops, onions, potatoes, rice and sugar-beets, as well as for con-

trol of snails in fish ponds and algae in rice-fields. Although triphenyltin
hydroxide is not licensed for general use in the U.S.A., its effectiveness in
protecting rice-fields has led to the Environmental Protection Agency authoris-
ing its emergency use in some parts of the U.S.A. as a fungicide on rice for a
period of one year [7]. Triphenyltin pesticides can be used in mixtures with
most of the commonly used insecticidal and fungicidal wettable powders. Mixtures with emulsifiable and oil-containing products should, however, be avoided.
These pesticides are said to have no harmful effects on treated crops provided
instructions for application are closely followed; indeed there are many reported cases of yields being improved following treatment, particularly with crops
of potatoes or sugar-beets. Triphenyltin hydroxide has also shown a repellant
effect on the army worm which attacks sugar-beet and other crops.

Triphenyltin chloride tends to be rather more phytotoxic than the other two
triphenyltin compounds and is used rather less in agricultural applications.
Yet it is very effective against fungi when its use can be tolerated.
"Tinmate", which is marketed by Nihon Nohyaku Co. Ltd. in Japan, contains 10%
triphenyltin chloride with inert ingredients and is recommended for use in controlling leaf spot and leaf blight on sugar-beet, late blight on potatoes and
Anthracnose and angular leaf spot on kidney beans. The manufacturer claims increased yield for sugar-beet treated with Tinmate. Sometimes the increased
phytotoxicity of triphenyltin chloride may be overcome by combining it with another pesticide, allowing lower doses of the organotin to be employed. Thus one
paper [8] describes the use of complexes of dimethyl sulphoxide or quinoline N-
oxide with triphenyltin chloride to protect tomatoes, celery and sugar-beet.

Following lengthy chronic toxicity testing, the World Health Organisation
pronounced the triphenyltin compounds (fentin acetate, hydroxide and chloride)
as "safe agricultural chemicals" and fixed an acceptable daily intake for man at
0-0.0005 mg/kg bodyweight. Table 6.3 shows recommended dosages and waiting
times (between treatment and harvesting) for the triphenyltins against a variety
of crop diseases.

The W.H.O. Report [9] considered that the chance of triphenyltin compounds
reaching the consumer via sprayed crops was very limited, since the residues of
these compounds or their metabolites occur only on the surfaces of those plant
parts which have been treated and are easily removed by peeling, washing, shell-
ing, etc., and are readily broken down in thermal processes such as cooking.

TABLE 6.3

W.H.O. recommendations for use of triphenyltin compounds to protect crops [9]

Crop	Disease	Recommended dosage kg a.i./ha	No. of applications	Safety period, days	Recommended tolerance ppm
potatoes	Phytophthora infestans of foliage and tubers, and Alternaria solari	0.18 – 0.4	1 – 10	7	0.1
sugar-beet	Cercospora beticola	0.18 – 0.4	1 – 3	14	0.2
celery and celeriac	Septoria apii	0.2 – 0.32 (in 600–1000 litre water)	1 – 7	21	celery 1 celeriac 0.1
carrots	Alternaria porri	0.2 – 0.32 (in 600–1000 litre water)	1 – 7	7	0.2
pecans	Fusicladium effusum, Gnomonia sp. and Mycospherella corvigen	0.22 – 0.45	2 – 8		
rice	various algae	0.75 – 1.0 (in 1000 litre water)			
groundnuts	Cercospora sp.	0.2 – 0.6		7	0.05
coffee	Colletotrichum coffeanum	0.2 – 0.6 (in >1000 litre water)			
cocoa	Phytophthora palmivora	0.2 – 0.6 (in 1000 litre water)			

188

Fig. 6.4. Spraying triphenyltin hydroxide on a smallholding in Kenya where coffee crops are grown.
Photograph courtesy: Coffee Board of Kenya.

6.3.2   Organotin miticides

Joint research between Dow Chemical Co., and M and T Chemicals Inc., revealed the marked acaricidal properties of tricyclohexyltin compounds. A patent in 1966 [10] claimed tricyclohexyltin compounds, including the hydroxide, halides, acetate, alkenyl and other more complex derivatives as pesticides for controlling arachnids. The toxic action stems from the tricyclohexyltin portion but the nature of the other group influences other properties; thus if water solubility is important, then the acetate is the preferred compound.

The hydroxide was adopted for commercial formulations; this is a white crystalline powder, almost odourless and with zero vapour pressure at 25°C. The compound has no true melting point but converts to bis(cyclohexyltin) oxide at 120-137°C and this oxide decomposes at 228°C. Technical material (95-96% tricyclohexyltin hydroxide) has an apparent melting point of 195-198°C. The compound is virtually insoluble in water and has a very low order of solubility in most organic solvents. Aqueous suspensions are stable over a pH range extending from slightly acid to alkaline; in strong acids, however, it reacts ionically to form salts. Reports of preliminary testing with the miticide have been summarised by Evans [11]. Extensive testing, both in the laboratory and in field tests in the U.S.A., Japan and Europe, demonstrated that formulations based on tricyclohexyltin hydroxide can give excellent control of motile forms of plant-feeding mites whether these are susceptible or resistant to organophosphates or other types of acaricide; some activity has also been noted against eggs. The compound has been assigned the short name "Cyhexatin" by the International Standards Organisation. Formulations based on cyhexatin are marketed under the trade name "Plictran" by the Dow Chemical Co. Plictran 25W is a wettable powder containing 25% tricyclohexytin hydroxide, whilst Plictran 50W contains 50% of the organotin. Plictran 600F is a flowable formulation containing 60% tricyclo-hexyltin hydroxide. Wettable powder formulations are pale bluish-grey in colour and are stated to have a minimum shelf life of 2 years. The formulations are readily dispersible in both soft and hard water. The W.H.O. has established the acceptable daily intake of cyhexatin for man at 0-0.007 mg/kg body weight, fol-lowing extensive short-term and long-term toxicity testing [12]. Table 6.4 gives dosage rates for Plictran on various crops, as recommended by the manu-facturer [13].

In contrast to the efficacy of tricyclohexyltin hydroxide against phytoph-agous mites, it has been shown to have little effect on most predacious mites and insects at recommended rates of application. Its use is thus complementary to the control of plant-feeding mites by biological means. A great deal of work has been performed by the manufacturers and by independent researchers to assess the environmental impact of the use of cyhexatin. Methods of analysis have been developed which are suitable for determining organic as well as total tin in wa-ter, soil, plants and animal tissues [9, 14, 15]. Typical residues which have been found on various crops in different countries after specified treatments are tabulated in Dow Chemical's technical publication [13]. The miticide has no systemic action, so that residues remain on the surface of treated crops. Plictran is now registered for use in 34 countries and is used widely on a mul-tiplicity of crops.

Sugavanam [16] has summarised work by Dow Chemical Co. on the environmental impact of tricyclohexytin hydroxide. When radio-labelled compound was fed to rats, 99.9% of the radioactivity was excreted indicating no significant absorp-tion in the gastro-intestinal tracts. The metabolism of the compound in animals occurs by stepwise cleavage of cyclohexyl groups from the tin atom:

$$(Cy)_3SnOH \rightarrow (Cy)_2SnO \rightarrow CySnO_2H \rightarrow Sn^{4+}$$

When Plictran was sprayed in three different locations over a period of 3 years, build-up in the top 6 inches of soil was negligible.

TABLE 6.4

Recommended dosage rates for Plictran [13]

Crop	Dosage	Application
apples pears peaches plums almonds apricots prunes strawberries blackberries raspberries grapes hazelnuts pecan nuts tea hops papaya	25-35 g a.i./100 litres	ground
Field and glasshouse crops ⁄ : tomatoes peppers beans cucumbers eggplants melons ornamentals + peanuts	0.35-0.4 kg a.i./ha or 25-35 g a.i./100 litres	ground
cotton	0.4-0.6 kg a.i./ha	ground
maize soybeans	0.4-0.7 kg a.i./ha	aerial

⁄ Some localised necrotic spotting of foliage or flowers has been observed following application to crops grown under glass. Addition of extra wetting agent
has been shown to increase the severity and so use of additional wetting agent,
or of tank mixes with other pesticides should be avoided under glasshouse conditions.

+ Application to ornamentals under glass is not recommended.

In 1969, an experimental acaricide, bis[tris(2-methyl-2-phenylpropyl)tin]
oxide, was synthesised by the Shell Development Co., California, under the code
number SD 14114. Following successful preliminary trials in the U.S.A. in 1969,
a more extensive programme of testing was implemented in 1971 which included
trials in all deciduous fruit- and citrus-growing areas, as well as studies with
grapes and nut crops. The findings, backed by results of trials in U.K., France
and Belgium, led to further development work in 1972 and 1973. In 1975 the com-

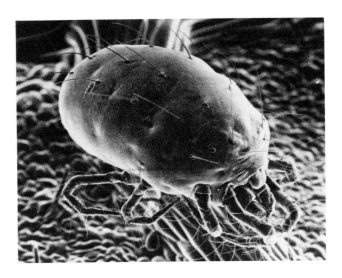

Fig. 6.5. Electron micrograph of a red spider mite.
Photograph courtesy: ICI p.l.c., Plant Protection Division, U.K.

pound was introduced commercially by Shell under the trade name "Vendex" in the
U.S.A. and as "Torque" outside that country.

The compound, more commonly known as fenbutatin oxide, is a white, odourless
solid melting at 145°C and is non-volatile (0.04% weight loss at 100 mmHg and
100°C). It is insoluble in water but is soluble in most organic solvents such
as xylene, benzene and methylene chloride. The compound kills adult and young
nymph forms of plant-feeding mites by contact activity and provides effective
residual control. Although it has only limited toxicity for eggs, the organotin
is persistent and even if the eggs are laid just before spraying, the mites will
be killed when they eventually hatch. The organotin is also active against
those species which have acquired resistance to other acaricides. Fenbutatin
oxide does not adversely affect the balance between phytophagous mites and their
predators; it also has minimum effect on beneficial insects such as bees and
ladybirds. No phytotoxicity has been reported over a wide range of crops and is
unlikely to occur, provided, the instructions for application are followed.

Torque is available as a 550 g/l suspension concentrate and as a 50% wettable
powder (often preferred for use in glasshouses), but other formulations may be
used, depending on individual requirements in various countries. The manufacturers state that one-pack formulations of Torque with other pesticides are also
available [17]. Dosage rates vary from 0.02 to 0.05% a.i., depending on the

crop, degree of infestation, time of year and temperature; the most common dose is 0.03% a.i. (i.e. 30 g Torque active material per 100 litre of spray). Table 6.5 shows recommended dosage rates for Torque on various crops. Since the formulations have been developed to give long-term control, effects may take 4 to 5 days to build up. Criteria such as the number of mites per leaf and the stage

TABLE 6.5

Recommended dosages of Torque on various crops [18]

Crop	Pest	Dosage rate, g a.i./100 litres water
Banana	Spider mite (Tetranychus lambi)	20
Citrus	Spider mites (T. telarius and T. kanzawai), Two-spotted spider mite	25-50
Fruit: berry, bush and cane	Two-spotted spider mite	45
Fruit: pome and stone	Red spider mite, Two-spotted spider mite	30-60
Glasshouse crops and nursery stock	Red spider mite, Two-spotted spider mite	25-50
Tea	Red crevice tea mite, Spider mite (T. kanzawai)	25-50
Vines	Red spider mite, Yellow mite	50

of crop growth are used as indicators of the time to spray. In general, treatment should be made before infestations reach a high level, i.e. not later than an infestation level of 5 mites per leaf.

This miticide is now approved and/or recommended for use in some 38 countries. It can be used on all commercially grown top-fruits including apples, pears, peaches, plums, lemons, limes, citron, grapefruit, oranges, vines, nuts and ornamentals (including roses, chrysanthemums, carnations, orchids and pot plants). It is also widely used as a range of glasshouse vegetables (cucumber, gherkin, melon, aubergines, tomatoes, red peppers and others).

In 1977 all available toxicological data for fenbutatin oxide were reviewed by the U.N. Food and Agriculture Organisation and the W.H.O. Joint Meeting of

Experts on Pesticide Residues. The Meeting concluded that results from short-term and long-term studies demonstrated no-effect levels and an acceptable daily intake for man was recommended as 0-0.03 mg/kg body weight [19].

The most recent organotin compound to be introduced commercially for agri-cultural use, is 1-tricyclohexylstannyl-1,2,4-triazole, now commonly known as "azocyclotin". This was first synthesised by Bayer A.G. in F. R. Germany in 1971 and after extensive testing, it was brought on to the market in 1980 under the trade name "Peropal". The compound occurs as colourless crystals which melt at 218.8°C; it is very sparingly soluble in common organic solvents. The X-ray crystal structure of azocyclotin reveals a polymeric chain structure in which near-planar cyclohexyltin groups are linked through N-1 and N-4 atoms of the triazole rings (see Fig. 6.6). The tin atom geometry is approximately trigonal bipyramidal [20].

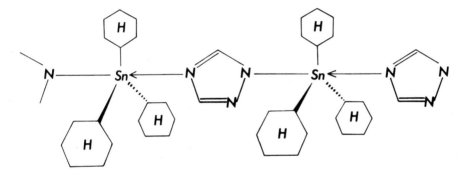

Fig. 6.6. Polymeric chain structure of azocyclotin [20].

The formulation is available as a 25% wettable powder and in some countries also as a 50% wettable powder. It is compatible with most common insecticides and fungicides. Peropal is stated to control all motile stages, including adult mites and larvae, of the most economically damaging species of spider mite. It has very good contact action and long duration of effectiveness, with rain re-sistance and excellent light stability [21]. It is harmless to bees and other beneficial insects and, like the other organotin miticides, exerts no systemic action. Greenhouse experiments showed Peropal to be effective against strains of Tetranychus urticae which had developed varying degrees of resistance to or-ganophosphate pesticides (see Table 6.6).

Peropal is recommended for use on pome and stone fruits, grapes, vegetables and citrus fruits where application rates of 0.1-0.12% of WP 25% are stated to give protection against the red spider and spider mites, including the fruit-

TABLE 6.6

Effectiveness of Peropal against resistant strains of T. urticae [20]

Strain of mite	Active ingredient %	Mortality, 2 days after application	Infestation level, 7 days after application
normally susceptible	0.1	100	0
	0.02	100	0
	0.004	100	0
	0.0008	99	0-1
	0.00016	70	3-5
	0.000032	0	5
moderately resistant to organophosphates	0.1	100	0
	0.02	100	0
	0.004	100	0
	0.0008	99	0-1
	0.00016	65	2-3
	0.000032	0	5
highly resistant to organophosphates	0.1	100	0
	0.02	100	0
	0.004	100	0
	0.0008	98	0-1
	0.00016	45	3-4
	0.000032	0	5

Infestation was rated on population density, 0 = no infestation, 5 = very severe infestation. Mortality is recorded for post-embryonic stages.

tree and spider mites, the two-spotted spider mite, the carmine spider mite and the citrus red mite. On bush and pole beans, applications of 0.6-1.2 kg/ha (depending on crop height) provide spider mite control. The manufacturers point out that special conditions in certain crop growing areas may necessitate amendment of these recommendations for application rates.

Toxicity assessments have shown Peropal to be acceptable for commercial use as an acaricide [20]. With regard to phytotoxicity, the pesticide is well tolerated by crops which normally require an acaricide to protect them.

6.3.3 Experimental organotin pesticides

As has been indicated earlier, the road from laboratory synthesis of a potentially useful pesticide to its commercial introduction, is a very long one and both the technical and the patent literature abound with details of new compounds and preliminary screening trials. A selection of these experimental biocides is reviewed here briefly, to illustrate current trends in organotin pesticide research.

One approach has been to combine a moiety of known pesticidal activity within an organotin compound. In one study [22] workers investigated the antimicrobial properties of some organometallic diethyldithiocarbamates against fungi causing red spot of sugar cane, root rot of sugar beet and leaf spot disease in sweet sorghum. Tributyltin, triphenyltin and tribenzyltin diethyldithiocarbamates were prepared and the activity was seen to increase tenfold when tributyltin replaced tribenzyltin. However, the biocidal activity was not as potent as might have been expected from a molecule containing two highly active components and this is attributed to the fact that most of the transition metal diethyldithiocarbamates, which are active biocides, possess a chelated structure, whereas the organotin derivatives have been shown to possess an ester-type structure.

Other workers in India [23] synthesised a series of isomeric chloro-oxazolylamines and their organotin derivatives and tested their fungicidal activity against *Piricularia oryzae*. All the organotin derivatives showed promising activity but those with a chlorine atom in the oxazole nucleus had the highest activity. It is interesting to note that maximum toxicity was obtained when the chlorine atom was directly linked to the oxazole ring rather than present as a substituent in the side chain. Kubo [24] attempted to combine trialkyltin and triaryltin moieties with phosphorus derivatives. These derivatives were the phosphordithioate, the phosphate, phosphinate and phosphorodiamidate. Antifungal and insecticidal activity of the compounds were assessed. All the compounds showed strong antifungal activity which was not very dependent on the type of phosphorus moiety. Only compounds of the type:

$$\begin{array}{c} RO \\ \diagdown \\ \diagup \\ RO \end{array} \overset{\overset{\displaystyle S}{\parallel}}{P}-S.SnR_3$$

showed insecticidal activity, correlating with increasing solubility in organic solvent. The phytotoxicity of the compounds varied somewhat; introduction of a diphenylphosphine grouping into the compounds considerably reduced the phytotoxicity. Phosphorus had no effect on the permeability of organotins into plant tissues.

In another study [25] the pesticidal properties of some organotin phenols, prepared from substituted phenols, were assessed against *Aspergillus niger* and *Chaetomium globosum*. The organotin phenolic compounds were considered to be potent multipurpose pesticides when they contained $3-CH_3$, $2-OCH_3$, $2-CH_3$ groups in the benzene ring. The 3-OH and 2-Cl substituents impart marked toxicity against fungi and $4-NH_2$ and $3-NO_2$ groups impart bacterial activity. Kourai et al. [26]

studied the antimicrobial activity of tripropyltin and tributyltin iso-carboxyl-ates and found that activity increased with a decrease in the number of carbon atoms in the main and side chains of the acyl groups. The activity of branched chain compounds was affected by the position of the branches.

Amongst some of the more interesting compounds described in the patent lit-erature, are tricyclohexyltin imide claimed as an active insecticide, miticide and lepidoptericide [27], monobutyltin triformate and monophenyltin triformate [28], a T.B.T.O./glucose complex [29] and tricyclopentyltin fluoride, claimed to have insecticidal and fungicidal activity with low toxicity [30]. Other organo-tin compounds which have been claimed, include di-n-butyl bis(diethyldithio-carbamate) stannane [31] and tri-2-norbornyltin compounds [32].

6.4 PRACTICAL ASPECTS OF ORGANOTINS IN CROP PROTECTION

6.4.1 Methods of application

All the organotin compounds currently in use as pesticides are virtually in-soluble in water and are sprayed as suspensions in water. The sprays may be made up from wettable powders which are first mixed to a paste with a little wa-ter and then poured with constant stirring into a spray tank half-filled with water. The remaining water is then added and the mixture agitated vigorously before application. In some cases, suspensions are supplied by the manufact-urers which may be diluted as needed, thus avoiding the need to handle dusty powders.

A range of spraying equipment is available, from simple hand-held syringes to machine sprayers, but they all operate on the same principle, which is a means of atomising the fluid and then propelling it through the air. The basic equip-ment consists of a tank to hold the treatment liquid, a pump to deliver the liq-uid to the nozzles at the required volume and pressure and one or more nozzles which will transform the wash into a fine spray. Knapsack sprayers are conven-ient for relatively small scale application, or a tank may be fitted with wheels and connected to a hand sprayer. For large-scale treatments, tanks may be mounted on vehicles with spray booms extending on either side of the vehicle (see Fig. 6.7).

Aerial spraying is sometimes employed but this has to be officially permitted for the pesticide and the crop. Maize and soybean crops may be protected by aerial spraying of Plictran and potato fields may be treated by aerial spraying of Du-Ter.

Fig. 6.7. Spraying fruit with Torque miticide.
Photograph courtesy: Shell Chemical Co., U.K.

Two types of spraying may be adopted in treating crops: high-volume spraying in which large amounts of a suspension with a lower concentration of toxicant are applied or low-volume spraying (discharging spray into an air blast) when a higher concentration spray is applied in smaller amounts. Thus Plictran 600F (a suspension of 60% tricyclohexyltin hydroxide in water) is recommended for application at not less than 560 litres of water per hectare in low-volume sprays and at least 2250 litres per hectare in high-volume sprays, at appropriate concentrations.

The timing and the number of applications are important and manufacturers' recommendations should be followed. On the whole, spraying in warm weather is more effective, since the biocidal activity of the organotin in general increases with temperature. Since the organotins act as contact poisons, it is essential to achieve complete and uniform coverage and care should be taken to spray flat surfaces and uprights.

Care should be taken when handling all pesticides, including organotin-based formulations. Manufacturers' instructions should be clearly read before spraying commences. Eye protection should be worn when pouring from the container and care should be taken not to inhale the spray mist which may develop when preparing the spray mixture. Body contamination (skin as well as mucous membranes) should be avoided, clothes should be changed after work and the operative should wash thoroughly. Empty containers are best disposed of by burning or by burying them in waste land away from water supplies. Care should also be taken when cleaning equipment after spraying.

6.4.2   Use of triphenyltins on potatoes

An important application of triphenyltin pesticides is to control leaf- and tuber blight on potatoes. Late and early blight (Phytophthora infestans and Alternaria solari) are two dreaded diseases in the potato crop. Late blight (leaf blight) is especially capable of epidemic appearance, the reason being its short incubation period (3-5 days), great capacity of the fungus for reproduction and the fact that the reproductive organs of the micro-organisms are immediately infectious in the presence of water. Within a few days this organism can cause complete defoliation leading to a considerable loss of yield during the period of major growth. This same organism is also responsible for tuber blight, the spores being washed down into the soil by rain. It results in much lower quality in the crop; when infected potatoes are stored, further losses occur. Tuber blight, scarcely detectable from the outside, is especially damaging to the seed potatoes because each infected tuber can cause the primary infection of a whole field, soon threatening large-scale crops. The Alternaria fungus requires a warm, dry climate with low rainfall. Although the blight caused by this organism does not have the economic consequences of late blight, it can still cause serious damage to leaves and stems, as well as tubers. Attack usually sets in before the blossom, so that control has to start early.

For effective protection, the fungicide should be sprayed as soon as weather conditions are favourable to infection, during the preflowering stage, and continuing throughout the season at 7-14 day intervals or at 7-10 day intervals during severe blight conditions. Sometimes application of the triphenyltin compound may be combined with use of other fungicides such as a dithiocarbamate, by spraying first with this agent and then switching to the organotin formulation as the season progresses. Triphenyltins effectively control leaf blights and kill spores washed into the soil, thus preventing tuber blight. The F.A.O./ W.H.O. recommend 7 days between final spraying and harvesting. Increased yields following treatment with triphenyltins have been reported by a number of workers [5].

6.4.3 Control of Cercospora leaf spot and powdery mildew on sugar beet

Triphenyltin compounds are effective against Cercospora beticola which causes leaf spot on sugar beets and also against powdery mildew. Leaf spot is readily identified by the presence of small brownish spots with reddish-purple borders, which give the leaf a speckled appearance. As the disease advances, the spots enlarge and turn grey and eventually ragged holes appear and the leaf dies. High humidity and fairly warm temperatures are favourable for development of the disease and it is most economically significant in warm climates. The yield and the sugar content of the roots may be reduced and large damaged parts of the beet must be cut away before processing can occur. Powdery mildew, although not as economically important as leaf spot, is nevertheless causing increasing concern to sugar-beet growers since when it occurs it can have a damaging effect on yield. Applications of the triphenyltin compound should ideally be made before the diseases occur, but if disease has set in, then stronger concentrations should be used. Manufacturers of Du-Ter recommend that applications should commence in May and be repeated at 14-21 day intervals, with a maximum of 3 applications, the final one being as late in the season as possible. Leaf spot and powdery mildew are both attacked simultaneously and treated crops show higher yields and higher sugar contents. The organotin formulations are stated to be much more effective than either copper preparations or dithiocarbamates against leaf spot. F.A.O./W.H.O. recommend 14 days waiting time before harvesting, without grazing on, or feeding of, treated tops. The waiting time should be extended to 28 days if tops are to be used as fodder.

6.4.4 <u>Control of leaf spot on celery</u>

Considerable losses to celery crops are caused by leaf spot, <u>Septoria apii</u>.
The disease first appears as spots, lighter or darker than the foliage on which
they occur. These spots will spread and destroy the whole leaf under suitable
weather conditions. Spores initially located on the older and outer leaves will
rapidly spread to younger, growing leaves. Spraying regularly with triphenyltin
formulations will keep celery cultures free from fungi and produce much higher
tuber yields. The organotins also have a curative effect in cases where infection has begun. The F.A.O./W.H.O. waiting time is 21 days if the leaves are intended to be used for consumption.

6.4.5 <u>Protection of coffee crops</u>

Several of the diseases which affect coffee crops are susceptible to treatment with triphenyltin compounds. Coffee berry disease is caused by
<u>Colletotrichum coffeanum</u>; leaves are marred by brown irregular-shaped spots
which is not economically serious, but staining of the berries lowers their market value. The same fungus can also cause black berry disease which renders the
coffee berries hard and brittle. Coffee rust, <u>Hemileia vastatrix</u>, manifests itself as small yellow spots on the underside of leaves; these increase in size
and become covered with orange fungal spores. Eventually the rust spreads to
the upper surface of the leaves which become brown and damaged. Fruitspot, or
grey blotch (<u>Cercospora coffeicola</u>) forms large blotches on the upper side of

TABLE 6.7

Effects of fungicides on coffee berry disease [33]

Fungicide	% inhibition of spore germination at different fungicide concentrations						Disease control rating [+]
	1000 ppm	500 ppm	100 ppm	10 ppm	5 ppm	1 ppm	
Untreated	-	-	-	-	-	-	10.44
Chlorthalonil (75% W.P.)	100	100	100	100	99	5	16.60
Captafol (50% W.P.)	100	100	100	100	100	100	6.02
Benomyl (50% W.P.)	100	100	94	84	83	80	1.27
Triphenyltin acetate (50% W.P.)	100	100	100	100	100	99	2.06
Triphenyltin hydroxide (50% W.P.)	100	100	100	100	100	88	1.99

[+] Field tests: Smaller the value, better the control.

the berry and affected berries can give trouble during processing. These can
all be controlled by repeated applications of triphenyltin compounds. Table 6.7
compares the effectiveness of various fungicides against <u>Colletotrichum</u>
<u>coffeanum</u> as demonstrated in laboratory and field tests conducted in Tanzania
[33].

Fig. 6.8. Coffee is one of the principal crops in Kenya and organotins are used
in its protection.
Photograph courtesy: Coffee Board of Kenya.

Smith [34] has reported that Du-Ter has been recommended by the Coffee
Research Foundation in Kenya for the treatment of leaf rust and as a protectant
against the giant looper caterpillar which attacks the leaves of coffee plants.
Other triphenyltin formulations are also undergoing trials there.

6.4.6   Triphenyltin compounds to protect rice fields

Triphenyltin compounds are used against rice blast ("Imochi-disease", caused
by <u>Piricularia oryzae</u>), against sheath blight (<u>Pellicularia sasakii</u>) and against
leaf spots (<u>Helminthosporium oryzae</u>). The development of algal infestation in
rice fields, which can seriously affect crops, can also be controlled with these
fungicides. To combat the fungal infections, two to three prophylactic applications of the triphenyltin are recommended, first during the booting stage, followed by a second one during flowering. In long duration varieties, a third
spraying at the beginning of maturity is necessary. The F.A.O./W.H.O. has not
yet recommended a waiting time and manufacturers recommend 14 days until a fig-

ure is specified.

To treat algal infection, spraying should take place directly and uniformly over the paddy water covered with algae. The effect is visible within a few hours, but the paddy water should not be changed for a week. Precautions are necessary to avoid contamination of foodstuffs or drinking water. It should also be borne in mind that many species of fish are very susceptible to organotin compounds.

Some phytotoxicity has been experienced in a few cases with triphenyltin compounds, thus in one study [35] triphenyltin hydroxide, whilst controlling rice blast, did produce some brown spotting on Padma and Jaya varieties of rice. In the U.S.A. triphenyltin hydroxide has not been licensed for use on rice, but as an emergency measure, it was given interim approval in 1983 for use for one year [7].

6.4.7 Protection of other crops against fungal diseases

A number of other crops are routinely protected with triphenyltin compounds. Leaf-spot diseases of carrots caused by Alternaria dauci and Cercospora carotae are world-wide in distribution and commonly occur in conjunction with each other, Cercospora being more severe on young leaves and Alternaria more on old leaves. These blights lower the yield of carrots and make mechanised harvesting more difficult. The fungi can be controlled by one to three sprayings with triphenyltins. Purple blotch (Alternaria porri) and grey mould (Botrytis allii) are very destructive diseases of the onion crop. Triphenyltins are able to control these fungi; a wetting agent is recommended to ensure good adherence of the fungicide to the crop. The waiting time is 10 days. Pre-bloom spraying and post-bloom spraying is employed to protect beans against leaf spot (Colletotrichum lindemuthianum and leaf rust (Uromyces phaseoli). The waiting time is 21 days if pods are to be consumed.

To control primary infection with downy mildew (Pseudoperonospora humuli) on hops, the stocks should be treated in spring, immediately after uncovering and cutting, either by hand application, covering stocks and soil at the rate of 300 litres of spray suspension (0.1% Brestan 60) per 1000 stocks, or by treating individual stocks at 100 ml/stock. The spraying should be repeated at 150 ml/stock before new growths reach 30 cm height. It is also reported that infection with downy mildew can be controlled by treating the soil with a 0.05% suspension of triphenyltin compounds [36].

Triphenyltin hydroxide is used, particularly in the U.S.A., to control a complex of diseases known as Pecan scab, which affect pecan nut trees. All parts of the tree have to be covered and application begins at the pre-pollination stage when young leaves unfold, a second spraying being necessary when the nuts are forming. The spray applications may be repeated at 2 to 4 week intervals to maintain control.

6.4.8 Applications of organotin acaricides

Spider mites, tiny eight-legged acarine creatures, were formerly of only minor or secondary importance, but in the course of the last few decades, they have assumed the status of economically damaging pests for a wide variety of crops. Spider mites live on plant juices; they settle mainly on the leaves and green stem sections, puncture the plant tissue and suck out the cell contents. Each mite may make about 20 incisions per minute into leaf tissues, resulting in mechanical damage as well as a weakening effect due to removal of sap. The first sign of mite attack is a speckling of the leaves which eventually turn a rusty colour and dry up and fall. Mites will often over-winter in the egg stage on the spurs and bark of trees; hatching commences in April or early May, depending on the weather conditions. The complete egg-to-adult cycle takes about 14 days and an adult female lives for about three weeks, during which time she lays about 120 eggs. These hatch within 5 days, so that the mites multiply rapidly. Spider mites have more motility when temperatures exceed 24°C and thus it is in summer that they can spread most easily through a crop, lowering the vitality and yield of the plants. The increased incidence of mites has given rise to problems of control, intensified in turn by the acquired resistance of the phytophagous mites to conventional pesticides.

The most commonly encountered mites are the fruit tree red spider mite (Panonychus ulmi), two-spotted spider mite (Tetranychus urticae), carmine red spider mite (Tetranychus cinnabarius), citrus red mite (Panonychus citri) and the pacific spider mite (Tetranychus pacificus). The organotin acaricides are effective against these and a great many other phytophagous mites.

Pome fruits are particularly susceptible to attack by mites; stone fruits are also at risk. Spraying should begin when the mites commence to hatch from the winter eggs, normally at petal fall, but before bronzing occurs. The organotin miticides are non-systemic and only affect those mites with which they actually come into contact, so that good application is the key to effective control. Comparatively high spray volumes are needed and good coverage is essential (of the underside of leaves as well as of the top surfaces). Spray jets should be adjusted to give a regular pattern with even distribution. Citrus fruit trees are normally sprayed with miticides for protection, but in recent years, resis-

tance to conventional pesticides has become a problem. Even the leaves of citrus trees are prone to attack by spider mites, which can inflict substantial
damage. The organotin miticides are effective against these resistant strains
and are widely used for this purpose. Tricyclohexyltin hydroxide, however, is
suspected of some phytotoxicity with citrus trees and the manufacturers of
Plictran do not recommend its use for this purpose.

Spider mite infestation of grapes causes discolouration at the puncture
points; the leaves and shoots become stunted and heavy mite attack can cause
leaf drop, with the result that grapes do not develop fully or fail to mature.
Organotin miticides are again effective in this application, protecting the
grapes without impairing their quality. One product which has been used as a
preventive measure, for vines in Switzerland, consists of a formulation containing Fenbutatin oxide and a copper-based fungicide. In a typical season, eight
applications at a comparatively low dose rate are stated to have given very good
control of mites and fungi on vines [17].

Hops may also be protected by organotin miticides, spraying beginning in early summer after training, as soon as over-wintering mites are seen on the foliage. A second spray is applied three weeks later or when mite build-up is
noticed on the underside of new leaves. Young growths less than 2 m high should
not be sprayed before the end of May. Soft, tender growth, or crops subject to
early season stress conditions may suffer slight leaf scorch which may be subsequently outgrown.

An interesting feature of the protection afforded by the organotin miticides
is the fact that maximum control is often achieved after an initial period of a
few days [37]. Even under conditions where they are slow to kill, they still
maintain protection, since mites will rarely feed on treated leaves. This
"antifeedant" effect of organotin pesticides is described more fully in 6.6.1.

6.5   EXPERIMENTAL STUDIES ON ORGANOTIN PESTICIDES

6.5.1   Fungicidal studies

In addition to the field and laboratory studies undertaken by the manufacturers of organotin pesticides, researchers all over the world are attempting to
find ways of developing their use, particularly in exploiting their biocidal
characteristics to protect other essential crops, perhaps in situations where
conventional pesticides have failed. These studies range from simple laboratory
screenings to full scale field trials and often factors such as costs and possible phytotoxic effects are taken into account. Some of the more interesting
or significant papers are reviewed here.

A number of workers have looked at the basic mechanisms by which these compounds exert their action. Knowles [37], in a study of the chemistry and toxicology of various acaricides, ascribed the mode of action of tricyclohexyltin hydroxide at least in part to an inhibition of oxidative phosphorylation. Both triphenyltin hydroxide and tricyclohexyltin hydroxide inhibit ATP ase in mice and in house flies. Tricyclohexyltin hydroxide is considered to be an outstanding inhibitor of oligomycin-sensitive (mitochondrial) Mg^{2+} ATP ase from fish brain and spider mite homogenates. Moore et al. [38] studied the effects of dibutylchloromethyltin on plant mitochondria. Workers had shown that lipoic acid plays a key role in the oxidative phosphorylation in mitochondria and Escherichia coli by functioning as the link between the respiratory chain and ATP synthesis. Dibutylchloromethyltin binds covalently to lipoic acid in submitochondrial particles and ATP synthetase complexes. Evidence is presented to show that the organotin is a potent inhibitor of the electron-transport chain in potato and mung bean mitochondria localised on the substrate side of the ubiquinone-cytochrome b region of the respiratory chain.

Singh and Mehrota [39] looked at the influence of fungicides, including triphenyltin acetate, on mycelial growth and respiration of Helminthosporium sativum, responsible for root and foot rot in wheat. No definite correlation was found between growth and respiration, but fungicides such as the triphenyltin, which inhibited the pathogen by 70% or more, also significantly reduced the rate of respiration. A number of workers have been concerned with developing effective fungicides to protect wheat crops. In one such study [40], 10 fungicides, including triphenyltin acetate, were screened against Drecholeva sativum which causes post-emergence rot in wheat in the seedling stage and leaf spot in older plants. Based on observation of inhibition zones of the fungus after incubating in petri dishes, triphenyltin acetate was the most effective compound, even at concentrations as low as 0.005%.

Seed-borne teliospores can set up karnal bunt of wheat and the possibility of using chemical seed treatments to control this disease has been studied by Rai and Singh [41]. The fungicides included triphenyltin acetate, hydroxide and chloride used as slurry treatments on infected seed. Seed treatment with triphenyltin hydroxide produced maximum inhibition (90-95%) of promycelial formation by teliospores for more than 67 days. The other triphenyltin compounds were effective for up to 23 days. However, none of the fungicides eliminated the seed-borne inoculum completely. Leaf rust of wheat, caused by Puccinia recondita is a commonly occurring disease under certain climatic conditions and reduces the yield and quality of grain. A number of fungicides have been screened in the laboratory against this fungus and triphenyltin chloride gave

100% inhibition at 0.1% concentration [42].

Tyagi and Singh [43] carried out laboratory and greenhouse evaluations of fungicides, including triphenyltin chloride, against Alternaria triticina which causes leaf blight in wheat. This is a very prevalent disease of wheat in many parts of India. Highly susceptible plants were given pre-inoculation sprays when plants were in the boot stage and then inoculated with a spore suspension after 48 hours and incubated for 60 hours before moving to a greenhouse. In the laboratory tests triphenyltin chloride was amongst the fungicides which considerably inhibited spore germination but it did not perform as well in the greenhouse tests.

Yildiz and Delen [44] studied the effects of fungicides, including triphenyltin acetate, on P. capsici which was responsible for disease in pepper plantations in Turkey. The organotin was active against P. capsici in these studies. It is interesting to note that the authors refer to other work in Turkey which had indicated that the organotin was translocated to the fruits in pepper plants, the amount found correlating with the number of treatments which the plant received. Triphenyltin compounds are not known to translocate in plants and exert no systemic action.

Krishnamurty et al. [45] tested fungicides and nematocides, including triphenyltin acetate, against Orobanche cernua which causes debilitating broom rape on tobacco plants. In a bio-assay test, the triphenyltin completely inhibited germination of the Orobanche seed; in a tobacco bio-assay test, pots treated with the organotin showed the lowest number of Orobanche shoots (less than 5 per pot). Other workers [46] showed triphenyltin acetate to be effective against seed-borne mycoflora affecting the sunflower crop; this is counted amongst one of the most important oil seed crops in India.

In the U.S.A., Walters [47] reported on an extensive study of the use of fungicides to control foliar, pod and stem diseases on soybeans in Arkansas. The most important foliar disease is Cercospora kikuchii; first symptoms are light purpling of the exposed leaves during the initial-seed and full-seed stages. Later, reddish-purple lesions appear on the upper and lower leaf surfaces, which may coalesce to form large necrotic areas and veinal necrosis. The most obvious result is the premature blighting and death of the young, upper leaves over large areas, or even entire fields. Under favourable conditions of high temperature and humidity, the disease progresses downwards, killing the entire plant. Rapid killing of the leaf prevents supply of food to the pods, resulting in reduced seed size and yield. Fungicides tested by Walters included triphenyltin acetate and they were applied by hand spraying or by aerial spraying (not more than 2.5 m above the tops of soybean plants) and by mist blowing. Seed was weighed after harvesting and its quality assessed by germination tests. Four-year average yield increases were obtained with several fungicides, including the organotin. The most economically significant benefits occurred when fungicides were applied to the soybeans at the beginning of seed growth and 14-21 days later. Triphenyltin acetate has not yet received official approval for general use on soybeans in the U.S.A.

Tikka disease accounts for at least half the loss in yield of ground nuts and 4 sprays at 10-day intervals with triphenyltin hydroxide (0.15%) from the 35th day, have been suggested for effective control. Following reports of some slight phytotoxicity at this concentration, workers in India [48] carried out field tests with various concentrations (0.075-0.15%) of triphenyltin acetate to determine the limiting concentration for effective protection. It was suggested, on the basis of the results, that 3 sprayings at 0.1% from the 40th day of the crop would give effective protection without scorching of the leaves. Attempts have been reported [49] to control the soil-borne pathogen Rhizoctonia bactaticola which affects large numbers of forest plants. The fungicides tested included triphenyltin acetate and whilst the organotin did not inhibit growth completely, it did minimise growth and field trials are to follow these laboratory studies.

6.5.2 Insecticidal studies

Some interesting synergistic effects have been observed during studies of the action of triphenyltin compounds on the cotton leafworm Spodoptera littoralis (Boisd.). In view of suggestions that the insecticidal action of trialkyltins may result from interference with ATP ase activity or related processes associated with oxidative phosphorylation, Kansouh [50] studied the relation between toxicity to S. littoralis and the carbohydrate content of treated cotton leaves. Five dilutions of triphenyltin hydroxide were used, in water or in a 2% aqueous solution of cane molasses and the mortality values against S. littoralis determined. The mixture of triphenyltin and molasses had double the toxicity of the organotin alone, based on LC_{50}; both were equally toxic on the basis of LC_{98} values. This is considered to indicate possible interaction between triphenyltin hydroxide and carbohydrate metabolism.

Other workers [51] studied the effect of adding synergistic agents to triphenyltin compounds or to tricyclohexyltin hydroxide, to increase their effectiveness against the cotton leafworm. The synergists, piperonyl butoxide, piperonyl sulphoxide and MGK 264 (proprietary formulation) were non-toxic when

used alone at the dosages involved in synergistic mixtures, but gave a high level of synergism with tricyclohexyltin hydroxide and low-to-moderate levels of
synergism with the triphenyltins. The optimum ratio of organotin to synergist
was 10 : 1; higher or lower ratios were less effective. It is hypothesised that
the synergist acts by inhibiting the detoxification of the insecticide in the
insect's body, reducing the rate of metabolism and prolonging the action of the
insecticide. At higher levels of synergist, penetration of the organotin may be
delayed or prevented. Two strains of leafworm were used, a susceptible laboratory strain (S) and a more tolerant strain (F). Toxicity of the various organotins and synergists was different towards the two strains. The F strain was
considered to possess a resistance to some insecticides due to some detoxification mechanisms which are very sensitive to organotin compounds.

Siddaramaiah et al. [52] showed that triphenyltin acetate was effective
against Beauveria bassiana which causes Muscardine disease in silkworms. The
infected worms become converted to white, stiff, caliciated bodies, covered with
fungal mycelium and spores, causing considerable losses to farmers. Treatment
with the triphenyltin, even at concentrations as low as 0.025%, inhibited fungal
growth.

Another study [53] showed up a paradoxical concentration effect in the toxicity of triphenyltin acetate towards insects. It had been found that under specific conditions, certain organotins may be biologically more active at lower
than at higher concentrations. Acetone solutions of triphenyltin acetate were
evaporated on glass and their toxicity assessed towards house-flies and larvae
of S. littoralis. As the concentration was increased, so the toxic effects decreased and this is ascribed to a polymerisation of the triphenyltin taking
place in acetone above a certain concentration, thus depleting the amount of
biologically active monomer.

6.5.3   Studies in mite control

The Yew podocarpus is one of Florida's most common landscape plants and it is
used for hedges, espaliers, sheared specimens and container plantings. Branches
and leaves are also used for florist greens in floral arrangements. A host-spe-
cific mite which has only been reported in Florida is Paracalacarus podocarpi;
its effects include shortened internodes and leaves producing typical rosetting
and tufting of infested terminals. When mite populations are high, they appear
as rust-coloured dust on infested leaves. To evaluate the effectiveness of 24
miticides, including tricyclohexyltin hydroxide and fenbutatin oxide, against
this species, infested plants were sprayed to the point of run-off, with a 7-day
interval between sprayings. Mite populations were evaluated at the start, after

7 days and after 14 days. Phytotoxicity was assessed at 3 days after each spraying and 21 days after the final application. Excellent results were obtained for the organotin miticides (see Table 6.8) with no apparent phytotoxicity [54].

TABLE 6.8

Effect of organotin miticides on P. podocarpi

Organotin	Concentration g a.i./litre	P. podocarpi, mean numbers		
		Before treatment	After 7 days	After 14 days
Tricyclohexyltin hydroxide 50% W.P.	0.3	354.1 ⨏	0	0.5
Fenbutatin oxide 50% W.P.	0.3	375.9 ⨏	0.4	0

⨏ Two 2-leaf samples taken per plant.

Some Soviet workers have also reported trials with organotin miticides. Thus one group found that spraying citrus trees with 0.2% fenbutatin oxide reduced the mite population by 99.1% after one month [55]. Murusidze et al. [56] found that spraying mandarin trees with 40% tricyclohexyltin hydroxide controlled red citrus and silvery mites, reducing the population by 95.3-97.3% and increasing the fruit yield from 13,720 to 19,580 kg/ha.

In Egypt, spider mites attacking cotton plants have become a serious problem and resistant strains have worsened the situation. In a search for new and effective specific acaricides for control of Tetranychus cinnabarius, some researchers there [57] conducted trials with a series of miticides, including tricyclohexyltin hydroxide. By the 7th day, 99.97% control was achieved with the organotin and residual effect studies showed that tricyclohexyltin hydroxide produced a 50% reduction in mite population 25 days after spraying.

A study by Muir and Cranham [58] illustrates one approach to determining the influence of pesticides on insects with acquired resistance. Damson-hop aphids were collected from cultivated sites in Britain and a susceptible stock obtained from wild hops remote from commercial spraying. Aphids were sprayed with serial concentrations of each insecticide under test. After treatment, the aphids were transferred to clean foliage so that only the contact action of the insecticides was measured.

Spider mites from hop gardens were cultured on dwarf French beans in cages in an insectary for at least one generation before testing. Standard susceptible and resistant strains were maintained on isolated leaf cultures which were used as nuclei for subsequent mass-rearing on beans in cages. Most bio-assays were performed on adult female mites using a modified taped-slide technique. No evidence was observed in these tests, of resistance to tricyclohexyltin hydroxide, the organotin showing equal effectiveness against susceptible and resistant strains.

In another study [59] the effectiveness of acaricides, including tricyclo-hexyltin hydroxide, against resistant and susceptible carmine spider mites was determined. Two colonies of T. cinnabarinus were raised on beans, in the laboratory. One colony came from roses which had received spray treatments and the other from a garden where no pesticides had been used. The strains were tested both in the laboratory and in field tests; the organotin killed more than 90% of resistant mites and was effective in the field. Multi-resistance is commonly found in various spider mite species on several crops. The time required to establish resistance to a particular pesticide can vary considerably even within one mite species on a particular crop. Since reversion to susceptibility is very slow in spider mites, prospects for a rotational spraying schedule are not very good. The authors urge workers to establish dosage-mortality data for all promising pesticides before resistance develops. It is now an essential part of modern pest control procedures to monitor continuously the level of pesticide resistance in major pests.

6.6   NEW APPROACHES TO PEST CONTROL

Chemical plant protection agents are an essential part of modern agriculture and provided the chemicals and their modes of application are carefully controlled, they constitute no environmental hazard. This is particularly true of the organotin pesticides which eventually break down into safe non-toxic inorganic forms of tin. However, modern pest management systems are aimed at keeping the use of toxic chemicals to a minimum and organotins offer promise for integration into these systems so that they can operate at maximum effectiveness. In particular, the organotins can be used to disrupt normal maturation processes and to disturb behaviour patterns, notably by inhibiting feeding.

6.6.1   Organotins as antifeedants

An interesting concept in pest control is the idea of using chemicals to inhibit feeding, i.e. antifeedants. Control against insects in the field was first encountered incidentally, by Murbach and Corbaz [60] with the Colorado beetle, Leptinotarsa decemlineata and by Solel [61] with the Egyptian cotton

leafworm, _Spodoptera littoralis_. Ascher [62] first established that the cause
was, in fact, an induced antifeedant effect, in experiments with sugar-beet
leaves treated with triphenyltin acetate on which larvae of _S. littoralis_
attempted to feed.

The use of anti-feedants in crop protection has several advantages over conventional techniques, the main one being that beneficial and non-target insects
are not affected because they do not feed on treated crops. The pest insect is
not killed but starves to death or is eaten by predators. This selectivity of
action means that antifeedants can be integrated with other pest control methods
such as trapping, mass sterilisation or the use of bacteria and viruses. A second advantage is that antifeedants will act sooner than conventional pesticides
to restrict feeding damage. During the time that a conventional insecticide
takes to act, the insect normally continues feeding. So far, the greatest limitation in the use of organotin antifeedants is the fact that no systemic compound has been found which would protect plants against sucking insects (such as
aphids) and internal leaf eaters, in addition to those surface-feeding insects
which are inhibited.

Haynes [63] has studied the mode of action of organotin antifeedants and has
reviewed work in this field. A true antifeedant disturbs the feeding of an insect at the level of the sense organs. The feeding behaviour of phytophagous
insects such as lepidopterous larvae can be regarded as a sequence of components
including oriented movements towards the food, inhibition of locomotion, test-
biting of the plant and finally the initiation and maintenance of continuous
feeding. Chemicals may act at any point in this sequence of activities, those
termed antifeedants acting to deter an insect from further feeding after a test
bite and from beginning an uninterrupted feeding phase.

The three conditions which must interact to sustain continuous feeding, are
that the insect must be hungry, must receive a gustatory stimulus from the intended food and must not detect any inhibitory signs. An antifeedant prevents
the insect from recognising its normal host plant gustatory stimulus by inhibiting taste receptors. In caterpillars, the contact chemoreceptors involved in
taste are situated on the mouth parts on the ventral side of the head. A number
of caterpillars also have a non-specific receptor which is sensitive to feeding
deterrents. In certain species, therefore, a chemical inhibitor of feeding may
act by stimulating a deterrent receptor.

There is however, some evidence that organotins do not act by disrupting sen-
sory input from the mouth parts. Findlay [63] showed that mortality in first

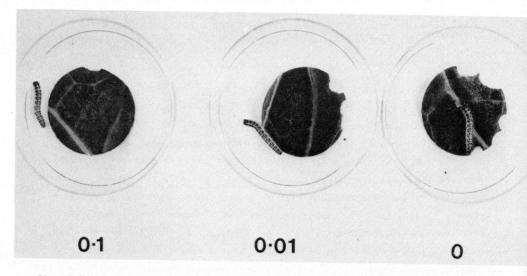

0·1   0·01   0

Fig. 6.9. A laboratory bio-assay of fourth instar Pieris brassicae larvae on cabbage leaf discs treated with triphenyltin acetate (dipped in 0, 0.01 and 0.1% solutions before testing). Note the increase in amount of leaf eaten as concentration of the organotin antifeedant decreases.
Photograph courtesy: Susan Haynes, Imperial College at Silwood Park.

and third instar larvae of S. littoralis tested with triphenyltin hydroxide and acetate, was due to stomach poisoning and not to an antifeedant effect. Ascher and Ishaya [64] showed the effect in S. littoralis larvae to be due to inhibition of the digestive enzymes protease and amylase in the larval gut. It can be seen that organotin compounds act as acute toxicants above certain concentration thresholds and it remains to be established whether or not the antifeedant effect is separate from, or a sublethal expression of, the toxic effect.

Numerous field trials have taken place with organotin antifeedants. In Kenya, it is reported that trials have been in progress for two years at the Coffee Research Foundation with the antifeedant effect of triphenyltin hydroxide on the giant looper caterpillar which attacks the foliage of coffee trees [34]. Other researchers have studied the persistence of protection against Spodoptera litura on castor plants, afforded by triphenyltins. The organotins were applied to leaves at concentrations of 0.05 and 0.025% using knapsack sprayers and leaves plucked 1, 2, 5, 10, 15 and 20 days after spraying were exposed to fourth instar larvae of S. litura. Persistent protection ($> 50\%$) was obtained over 10 to 15 days from one spraying, the higher dose giving longer protection to the leaves [65].

Chibber [66] evaluated the activity of triphenyltin compounds against
Diacrisia obliqua on sugar-beet. Sugar-beet was sown in 12 m^2 plots and sprayed, using a high-volume foot pump sprayer, with triphenyltin hydroxide (0.02 and
0.04%) and with triphenyltin chloride (0.045 and 0.09%). The leaves were sprayed again after 10 and after 23 days. Triphenyltin chloride at 0.09% provided
maximum protection 13 days after the second spray. Protection was significant
13 days after the second spray with triphenyltin hydroxide at both concentrations and with triphenyltin chloride at 0.045% concentration. Root yield was
also higher in treated plots.

Ascher [67] has studied the insect antifeedant effect of azacyclotin. Cotton
leaf discs with 24-hour old residues from dipping in aqueous suspensions of azacyclotin 25% W.P. were exposed for 48 hours at 27°C to 170-190 mg larvae of
S. littoralis in petri dishes. Leaf protection and larval starvation were determined at different concentrations and 50 and 95% protection (PC_{50} and PC_{95})
and 50 and 95% starvation (SC_{50} and SC_{95}) concentrations were determined by re-
gression analysis. The values obtained: PC_{50} 0.016%, PC_{95} 0.12%, SC_{50} 0.013%,
SC_{95} 0.11%, indicated considerable antifeedant effect.

6.6.2   Organotins as chemosterilants

Another approach to pest management is by interfering with the reproductive
cycle or the normal maturation process by chemical means. The organotin pesticides have been reported as having some sterilising effects against mites but
this approach has not yet achieved practical importance.

Salem et al. [68] investigated the capability of triphenyltin hydroxide for
inhibiting reproduction of the spiny bollworm, Earias insulana. The organotin
was dispersed in 20% honey solution at concentrations from 0.062 to 0.50%.
Pairs of newly emerged moths were placed in glass jars covered with wire screens
and kept in the dark. Okra pods were present as oviposition sites and a small
plastic vial contained cotton wool soaked in the honey solution. After 24 hours
this was replaced by normal untreated honey. The number of deposited eggs was
counted daily and the percentage of mated pairs was established by dissecting
the females after death. Ovarioles and testes were measured 2, 4 and 8 days after treatment. The organotin was shown to have a significant effect on the number of eggs laid and on the reproductive capacity of the female moths. The
higher the concentration of organotin used, the greater was the effect. The
oviposition period in the female was rendered much shorter. However, the percentage of hatched eggs was subject to considerable fluctuations and as time
passed, the number increased i.e. sterility was only temporary. The organotin
had only a slight effect on mating frequency. There was a reduction in the num-

ber of mature eggs and in the percentage maturation of eggs after treatment with
triphenyltin hydroxide.

Also in Egypt, workers studied the effects of triphenyltins on the red flour
beetle, Tribolium castaneum. Concentrations of 10 to 100 ppm of the organotin
were added to wheat and the influence studied on insect fecundity, fertility and
duration of the larval and pupal stages of T. castaneum. Triphenyltin hydroxide
was more effective than the acetate on adults and larvae. Adults were more susceptible than larvae to both insecticides; the LC_{50} decreased as the period of
exposure increased. The organotins were considered to have significantly affected the fecundity and fertility of insects treated with sublethal doses, particularly of the hydroxide. Both compounds prolonged larval-pupal durations.

6.6.3   Integrated control schemes

The trend in modern pest management is towards integrated control schemes in
which several approaches are combined. The special characteristics of the organotins: selective action, no tendency to acquired resistance by target species,
protective antifeedant activity, make them well-suited to fit into such schemes.
One such scheme, involving the use of tricyclohexyltin hydroxide to increase
yields of strawberry crops, has been described [70]. Strawberries are cultivated in Southern California as an annual crop on fumigated land. Both winter
(planted in November) and summer (planted in August) plantings provide continual
harvests from February to June. When these culturing practices are adopted, the
two-spotted spider mite, Tetranychus urticae and the strawberry aphid,
Chaetosiphon fragaefalin are consistently the only serious pests of this crop.
One possible control measure is the mass release of predatory mites to suppress
T. urticae. Currently, control of this pest is achieved by foliar acaricide
sprays, usually applied when mite densities are low. Although tricyclohexyltin
hydroxide is "pest-specific", its frequent use at low spider mite densities
would not only delay establishment of natural control, but would also increase
the possibility of acquired resistance by the pests. The authors report that
over the period 1976 to 1978, they have been studying the effect of differing
levels of T. urticae populations on strawberry yield, with a view to establishing a threshold level of infestation which is still economically acceptable.
Use of a threshold treatment would delay the onset of acaricide resistance and
enhance the establishment of an integrated pest control management system.
Varying infestation levels were established (by applying the organotin at 0-5,
25, 50, 75 and 100 mites per leaflet pest density) and the subsequent strawberry
yield responses determined. Results showed that establishment of an infestation
level at 50 mites/leaflet would provide effective control. Seven predatory species were present and tricyclohexyltin hydroxide had no effect on these, the

predator density depending largely on host density.

6.6.4   Slow release systems

Interest has been shown in recent years in the use of slow release systems as a means of dispensing the minimum amount of a biocide into the environment for effective control. The concept is already employed in antifouling systems and in some medical applications and there is no reason why the principle should not be applicable to crop protection, with long-term release of small doses taking the place of large doses applied in one treatment to destroy target organisms before the toxicant is lost. This would not only be economical in the amount of pesticide used, but would also create less burden on the environment. Cramer [71] has discussed the potential of controlled release pesticides. He described the concept of a formulation containing a conventional "free" and a "bound" component. The first would be applied at the minimum concentration needed to inhibit the pest population immediately (MIC); the second would be bound into a form that could not be degraded by the environment and would be released at the same rate as the first component degraded. This would ensure that the chemical concentration and the pesticidal activity would be stationary at the MIC level or just above. The amount of bound component would determine the duration of the pesticidal activity. Cramer goes on to illustrate how controlled release formulations could reduce environmental impact by more than 50%. Impregnation of the pesticide in rubber or plastic matrices could be one approach to slow release, the toxicant being released by a diffusion process.

To date, relatively little has been done in the application of controlled release methodology to agricultural chemicals. Most of the activity has been concerned with the development of bait formulations based upon the use of pheromones to lure the target pest into a mechanical or toxicant trap. Cardarelli [72] has reviewed the work that has been conducted with non-bait formulations. These include absorption of insecticides on lignocellulosic or proteinaceous waste materials or on carriers such as wood chips, corn cob dust, etc., coated with a microcrystalline wax emulsion. To date there have been no developments involving organotin pesticides; the fact that organotins do not impose a long-term pollution burden on the environment makes this approach rather less critical than in the case of more permanently toxic substances. No doubt, however, as controlled release systems develop, organotin pesticides will be incorporated in view of the savings in materials utilisation which are implied.

REFERENCES

1   B. Sugavanam, Tin and its Uses, 126 (1980) 4-6.
2   G.J.M. van der Kerk, 8th Ann. Symp. Crop Protection, Ghent, 1975, 35 pp.

216

3   G. Huber, Tin and its Uses, 113 (1977) 7-8.
4   Anon, W.H.O. Geneva, 1975, Tech. Rep. No. 560.
5   R. Bock, Residue Reviews 79 (1981) F.A. Gunther (Ed.), Publ. by Springer-Verlag, 1981, 270 pp.
6   K. Härtel, Tin and its Uses, 43 (1958) 9-14.
7   Emergency exemption granted under Section 18 of the Federal Insecticide, Fungicide and Rodenticide Act for a period of 1 year from June 3rd, 1983; private commun. to Tin Research Institute, Inc., Aug. 25th, 1983.
8   K.G. Kumar Das and Chan Kai Cheong, Planter, Kuala Lumpur, 51 (1975) 355-364.
9   F.A.O. and W.H.O., Rome, 1971, AEP : 1979/M/12/1, 327-366.
10 E.E. Kenaga, U.S. Patent 3,264,177 (1966).
11 C.J. Evans, Tin and its Uses, 86 (1970) 7-9.
12 W.H.O. Tech. Rep. Ser. 1974 No. 545, 440-452.
13 Plictran, The Dow Chemical Company Bull. EU 6550-6-1279, 12 pp.
14 H.B. Corbin, J. Assn. Offic. Anal. Chem., 53 (1970) 140-146.
15 F.A.O. and W.H.O., Rome, 1971, AEP : 1970/M/12/1, 521-542.
16 B. Sugavanam, Tin and its Uses, 126 (1980) 4-6.
17 Torque, Shell Chemical Co., Publicn. TOR/ENG/1/3.81.
18 Torque, A Shell Insecticide, Shell Chemical Co. Publicn. 4 pp.
19 F.A.O. and W.H.O., Rome, 1977, Evaluations of Some Pesticide Residues in Food, 1977, 229-261.
20 I. Hammann, Nachrichten Bayer, 31 (1978) 61-83.
21 Peropal, Bayer Agrochemicals Div. Report, E2-821/843304, 4 pp.
22 K. Singh, K.K. Bajpai, S.R. Misra, T.N. Srivastava and J.S. Gaur, Indian J. Exper. Biol. 12 (1974) 588-598.
23 B.K. Pattanayak, D.N. Rout, J.P. Nath and G.N. Mahapata, Indian J. Chem. 18B (1979) 286-289.
24 H. Kubo, Agr. Biol. Chem. 29 (1965) 43-55.
25 G. Mehrotra and A.N. Dey, J. Indian Chem. Soc. 52 (1975) 197-198.
26 H. Kourai, K. Takeichi and I. Shibasaki, J. Ferment. Technol., 50 (1972) 79-85.
27 Stauffer Chem. Co., U.S. Pat. 4,036,958 (1977).
28 VEB Electrokemisches Kombinat, Brit. Pat. 1,081,969 (1967).
29 East Ger. Pat. 71,661 (1970).
30 M.H. Gitlitz and J.E. Engelhart, U.S. Pat. 4,191,698 (1980).
31 Farbwerke Hoechst A.G., Brit. Pat. 835,546 (1960).
32 M & T Chemicals Inc., U.S. Pat. 3,781,316 (1973).
33 Fungicides, E. African Trop. Pesticides Res. Inst. Ann. Report, 1971, b.
34 F.E. Smith, Tin and its Uses, 126 (1980) 6.
35 K.V.S.R.K. Row and S.Y. Padmanabhan, Pesticides, 13 (1979) 27-30.
36 Codex Committee on Pesticide Residues; Fentin acetate, Fentin chloride, Fentin hydroxide, 5th Session, 28th Sept. - 6th Oct., 1970, The Hague, Netherlands.
37 C. Knowles, Environment. Health Perspectives, 14 (1976) 93-102.
38 A.L. Moore, P.E. Linnett and R.B. Beachey, Biochem. Soc. Trans., 7 (1975) 1120-1122.
39 K.B. Singh and R.S. Mehrota, J. Indian Bot. Soc., 57 (1978) 1-5.
40 S. Kulkarni, A.L. Siddaramaiah and K.S. Krishna Prasad, Curr. Res. (Univ. Agric. Sci.), 7 (1978) 51-52.
41 R.C. Rai and A. Singh, Seed Res., 7 (1979) 186-189.
42 N.L. Pleshney, N.N. Khune and P.G. Moghe, Pesticides, 14 (1980) 21-23.
43 P.D. Tyagi and M. Singh, Pesticides, 13 (1979) 35-36.
44 M. Yildiz and N. Delen, J. Turkish Phytopathol., 8 (1979) 29-39.
45 G.V.G. Krishnamurty, R. Lal and K. Nagarajan, Tobacco Res., 5 (1979) 89-92.
46 Z.S. Kanwar, P.K. Khanna and M.L. Thareja, Pesticides, 13 (1979) 41-44.
47 H. J. Walters, Univ. Arkansas, Agric. Exptal. Station, Rep. Ser. 250, Feb. 1980, 25 pp.
48 A.L. Siddaramaiah, R.V. Hiremath and K.S. Krishna, Indian J. Plant Prot., 5 (1977) 193-194.

49 A.L. Siddaramaiah, S. Kulkarni and A.B. Basavorajaiah, Pesticides, 14 (1980)
    25-26.
50 A.S.H. Kansouh, Bull. Ent. Soc. Egypt Econ. Ser. IX, 9 (1975) 349-354.
51 H.S.A. Radwan, M.R. Riskallah and I.A. El-keie, Toxicol., 14 (1979) 193-198.
52 A.L. Siddaramaiah, S. Lingaraju and K.S. Krishna Prasad, Indian J. Seri., 18
    (1979) 6-8.
53 K.R.S. Ascher and N.E. Nemny, Experientia, 32 (1976) 902-903.
54 J.A. Reinert, J. Econ. Entomol., 74 (1981) 85-87.
55 T.N. Novitskaya and N.V. Abzianidze, Subtrop. Kul't., 6 (1978) 76-77.
56 G.E. Murusidze, O.V. Sharashidze and L.T. Gogadze, Subtrop. Kul't., 6 (1978)
    78-80.
57 A.E. Salama and H.T. Farghaly, Bull. Ent. Soc. Egypt Econ. Ser. IX, 9 (1975)
    61-66.
58 R.C. Muir and J.E. Cranham, Proc. Brit. Crop Prot. Conf. Pests and Disease,
    1 (1979) 161-167.
59 F.A. Mansour and H.N. Plant, Phytoparasitica, 7 (1979) 185-193.
60 R. Murbach and R. Corbaz, Phytopath. Z., 47 (1963) 182-188.
61 Z. Solel, Israel J. Agric. Res., 14 (1964) 31.
62 K.R.S. Ascher and G. Rones, Int. Pest Control, 6 (1964) 6-8.
63 J.B.R. Findlay, Phytopath., 2 (1970) 91-96.
64 K.R.S. Ascher and I. Ishaya, Pestic. Biochem. Physiol., 3 (1973) 326-336.
65 A. Abdul Kaleem, S. Sodakathulla and T.R. Subramanian, Annali. Univ. Agric.
    Res. Ann. 4-5 (1974) 154-157.
66 R.C. Chibber, Indian J. Agric. Sci., 50 (1980) 176-178.
67 K.R.S. Ascher, Naturwissenschaften, 67(B) (1980) 312.
68 Y.S. Salem, M.I. Abdel-Megeed and Z.H. Zidan, Bull. Ent. Soc. Egypt Econ.
    Ser. IX, 9 (1979) 293-298.
69 A.K.M. El-Nahal, A. El-Halfawy and M.M. El-Attal, Bull. Ent. Soc. Egypt
    Econ. Ser. IX, 9 (1979) 31-36.
70 J.A. Wyman, E.R. Oatman and V. Voth, J. Econ. Entomol., 72 (1979) 747-753.
71 R.D. Cramer, Int. Pest Control, 1976, Mar./Apr., 4-8; May/June, 4-8.
72 N. Cardarelli, Controlled Release Pesticides Formulations, CRC Press, 1976,
    208 pp.

Chapter 7

MEDICAL USES

The actual or potential uses of organotins in the medical field can be divided into two areas: those where the organotin is used directly, for treatment as a therapeutic drug and those where the compound is used as a preventative measure. Although a great deal of research is being conducted in both these areas, actual usage of tin in medical applications is still very small. Nevertheless, there are developments which seem likely to lead to an increased use of tin in medicine in the not too distant future.

7.1 THERAPEUTIC DRUGS BASED ON ORGANOTINS

7.1.1 Anti-tumour drugs

One of the most outstanding recent developments in the field of metal compounds in medicine was the discovery, in 1969, of the anti-tumour activity of certain platinum complexes. One of the first, cis-dichlorodiamine platinum(II), commonly referred to as "cis-platin", showed activity comparable to existing organic drugs. A vast body of research led to a second generation of platinum complexes with a potentially higher therapeutic index. In the late 1970s, workers at the International Tin Research Institute in London began synthesising a series of diorganotin dihalide complexes, $R_2SnX_2L_2$ where R = methyl, ethyl, n-propyl, n-butyl or phenyl, X = chlorine, bromine, iodine or thiocyanate and L = oxygen- or nitrogen-donor ligand. These were modelled on the original active platinum compounds (Fig. 7.1).

The organotin compounds chosen for test contained cis-halogen groups, which seem to be an essential prerequisite for anti-tumour activity [1]. Some thirty octahedral diorganotin complexes were prepared and sent for testing to the Institut Jules Bordet in Brussels, under the auspices of the U.S. National Cancer Institute, in accordance with standard protocols for primary screening [2]. A number of these compounds have shown activity in vivo towards the P-388 lymphocytic leukaemia tumour in mice. Effectiveness is rated as the ratio of the median survival time (days) of treated mice (T) and untreated animals (C); a compound is considered to be active if T/C $\geqslant$ 120%. The compounds are generally of low water solubility and have been administered intra-peritoneally in saline, in saline and Tween 80 or as a suspension in saline. Typical results are shown in Table 7.1.

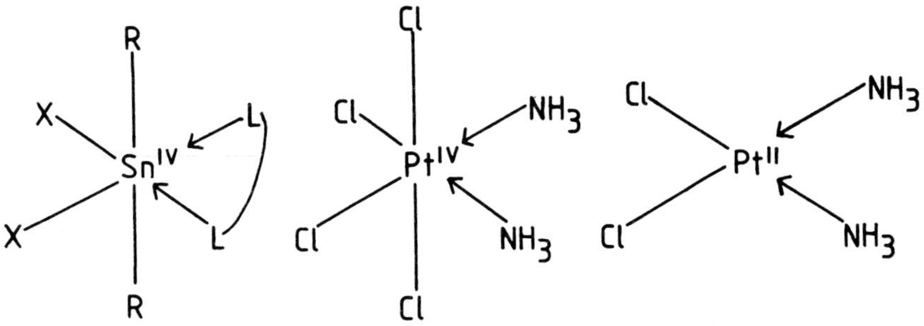

Fig. 7.1. Structures of platinum-based anti-tumour drugs and analogous tin(IV) compounds.

The organotin compounds tested to date have T/C values in the range 120-160%, which are lower than the values for the platinum compounds (200%), but on the other hand there is no evidence to suggest that the organotin compounds display a high nephrotoxicity (which is a problem with some of the platinum compounds).

The diethyltin complexes and diphenyltin dichloride .AMP showed the highest activity against the P-388 tumour system (AMP is 2-aminomethylpyridine). Since diethyltin dichloride is itself active (T/C = 125%) and the ligands are inactive, the mode of action is considered possibly to involve initial transportation of the complexed diethyltin dihalide compound into the tumorigenic cells, followed by reaction of the diethyltin dihalide (or one of its hydrolysis products) at the active sites. In line with this hypothesis, it was established that diethyltin oxide, $(Et_2SnO)_n$, the final hydrolysis product of the dichloride is also active against the same tumour system (T/C = 137%).

Workers had proposed that the anti-tumour activity of such complexes, as well as that of cis-platin, was dependent upon the Cl—M—Cl bond angle and hence the corresponding non-bonding Cl......Cl distance (bite). Only those compounds for which the Cl—M—Cl angle was < 95°, giving a "bite size" of < 3.6 Å (the upper limit for DNA-metal cross-links) would be active. An examination of crystallographic data for a large number of organotin complexes shows that the structure/activity relationships for these complexes does not follow this pattern [3].

TABLE 7.1

Diorganotin complexes which exhibit reproducible activity towards P-388
lymphocytic leukaemia in mice [1]

Complex			Dose (mg/kg)						
			400	200	100	50	25	12.5	6.25
Me_2SnX_2, L_2									
X = Br	L_2 = bipy		O	135	131	109	109	–	–
	L_2 = phen		O	O	O	131	122	115	117
X = I	L_2 = bipy		O	O	121	121	109	95	–
Et_2SnX_2, L_2									
X = Cl	L_2 = phen		O	O	152	148	139	131	–
X = Br	L_2 = phen		O	O	139	141	143	131	–
X = I	L_2 = phen		O	159	149	141	125	108	115
X = NCS	L_2 = bipy		O	O	O	O	152	157	145
	L_2 = phen		O	132	157	142	121	–	–
Pr_2SnX_2, L_2									
X = Cl	L_2 = phen		O	O	127	121	117	111	110
X = Br	L_2 = phen		O	O	O	115	116	114	–
Bu_2SnX_2, L_2									
X = Cl	L_2 = AMP		O	O	O	139	116	116	104
X = Cl	L_2 = phen		O	121	121	112	112	–	–
X = Cl	L_2 = bipy		130	117	112	109	–	–	–
X = NCS	L_2 = bipy		O	O	105	106	123	118	110
X = F	L_2 = phen		O	O	O	O	127	139	129
Ph_2SnX_2, L_2									
X = Cl	L_2 = AMP		O	O	124	148	153	O	–

From the data in Table 7.2 it can be seen that compounds with Sn—N bond lengths
greater than 2.390 Å show anti-tumour activity, whereas those with shorter bond
lengths are inactive. This suggests that the more stable complexes have lower
activities, which in turn implies that a pre-dissociation of the bidentate lig-
and may be a crucial step in the formation of a tin-DNA complex.

Over 100 diorganotin dihalide and dipseudohalide complexes with nitrogen

TABLE 7.2

Crystallographic and anti-tumour activity data for diorganotin dihalide
complexes [3]

Complex	Bond angle	Bond lengths Å		Activity (against P-388 tumour)
	Cl͡SnCl	Cl......Cl	Sn—N	
Ph₂SnCl₂.bipy	103.5°	3.94	2.359	103
Et₂SnCl₂.bipy	104.2°	4.00	3.375	115
Bu₂SnCl₂.bipy	104.3°	4.00	2.400	131
Et₂SnCl₂.phen	105.2°	4.02	0.413	176
Et₂SnCl₂.pdt ⁺	103.2°	3.88	2.50	144 *

⁺ 3-(2-pyridyl)-5,6-diphenyl-1,2,4-triazine.

* Presumptive activity.

donor ligands have now been synthesised at the Institute (Fig. 2) and tested for
anti-tumour activity; other types of structure currently being studied include
the di(hetero-aryltin) dihalide adducts.

Fig. 7.2. A large number of new organotin compounds are synthesised in the
laboratories of the International Tin Research Institute for screening as anti-
tumour agents.

Other workers have also been studying potential anti-tumour agents based on organotins; Sadler [4] has stated that up to August, 1980, some 1434 tin compounds had been tested by the National Cancer Institute in the U.S., of which 170 had shown some activity. A number of compounds have been patented, for example $(Et_2SnO)_n$, $Cl(Me_2Sn)_2O$ and $Ph_2Sn(OH)Cl$ [5] and diorganotin(II) derivatives of 5-fluoro-uracil [6]. Barbieri et al. [7] examined the anti-tumour properties of diorganotin(IV) complexes with adenine (R_2SnAd_2) and glycylglycine (R_2SnGly-Gly). The R_2SnAd_2 complexes were definitely active against the P-388 leukaemic tumour in mice, the T/C values being equivalent to those shown for the $R_2SnX_2.L_2$ series of adducts. The action of water on the R_2SnAd_2 complexes and the $R_2SnX_2.L_2$ adducts could yield analogous species, due to the possible co-ordination of H_2O to the metal centre, followed by the gradual dissociation of the Sn—N or Sn—X bonds due to hydrolysis. As a consequence, the same mechanism could be advanced for the anti-tumour action of the two series. This would consist of preferred transportation of the complex species into the tumour cells, which would be attacked by hydrolysed $R_2Sn(IV)$ moieties. Alternatively, it might be assumed that the facile hydrolytic cleavage of the co-ordinated N→Sn bonds, presumably taking place in the $R_2SnX_2.L_2$ adducts, does not occur to the same extent for the covalent N—Sn bonds in R_2SnAd_2 complexes, which would then act as anti-metabolites.

The trigonal bipyramidal R_2SnGly-Gly complexes showed general activity against leukaemia at very small doses with the butyltin, octyltin and phenyltin groups being the most active. It was interesting to note that these compounds all had comparable toxicity, whereas LD_{50} data for orally administered dialkyltin chlorides showed a progressive decrease in toxicity as the alkyl chain length increases. The authors therefore consider that the anti-leukaemic activity of R_2SnGly-Gly depends on the structure and the bonding. The Gly-Gly^{2-} moiety is thought to be effective in bringing the complex into the tumour cell, where it operates as an anti-metabolite.

Other workers [8] have reported on the anti-tumour activity of a broad range of organotin compounds. The most active was $[PhP(S)S]_2SnPh_2$ which had a T/C value of 140% at a dose of 12.5 mg/kg and of 130% at 6.25 mg/kg. Another interesting toxic effect of certain dialkyltin dihalides has been reported by Seinen [9] who found that when weanling rats were fed on diets containing di-n-octyltin dichloride, or di-n-butyltin dichloride, there was a decrease in the size of the thymus and thymus-related lymphoid organs. Once the organotins were removed from the diet, the thymus weight began to increase again. Thymus atrophy occurred at 5 ppm organotin and since water-soluble compounds had no effect, the atrophy is probably related to water-lipid partition of the compounds. Di-n-octyltin and di-n-butyltin chlorides exert a selective cytotoxic effect on lymphocytes in the thymus and in thymus-dependent areas of peripheral lymphoid organs without inducing a generalised toxicity or a myelotoxicity. The author considers this behaviour to offer possibilities for a therapeutic use of the compound in controlling various pathological events in which undesired T-lymphocyte functions are involved. Because of the selectivity of the action, the compounds may be useful as anti-T-cell tumour compounds.

7.1.2   Other possible uses of organotin-based drugs

In a series of papers, Kappas showed how the ability of tin protoporphyrin to inhibit the degradation of haeme to bile pigment could be put to therapeutic use. Haeme, which derives from haemoglobin and various cellular haemoproteins, is degraded to biliverdin by the microsomal enzyme haeme oxygenase. Strangely, this oxidation is induced by inorganic tin. Treatment with stannous chloride increased the rate of haeme oxidation in rat kidney by 20-30 times over a control [10]. Tin protoporphyrin, in contrast, has a much greater affinity for the catalytic site on the enzyme than does the substrate haeme and bile pigment formation is diminished by 98% in the presence of this compound [11]. Such a reaction has been proposed as a method of preventing hyperbilirubinaemia in newborn infants. Bilirubin is formed from haeme in a two-step enzyme reaction and this is a potential toxin for the central nervous system for infants in the period immediately after birth when the blood-brain barrier is still permeable to many substances. Previous chemical treatments have been largely ineffective, whilst therapies such as bathing in fluorescent light to break down the bilirubin, or partial blood transfusion, have undesirable side effects or involve possible risks.

Kappas and his colleagues [12] conducted trials using rats with tin protoporphyrin and other metallo-protoporphyrins. On administration of the tin compound, bile pigment formation diminished and serum bilirubin levels dropped rapidly, the effects lasting for several weeks, depending on the dose. The lowest effective dosage was established at 10 μmole/kg, which, when fed to new-born rats, produced low levels of enzyme activity over the 14-day study and also caused an immediate lowering of renal haeme oxygenase activity and a halt in the developmental increase of splenic haemic oxygenase activity. Within 24 hours, serum bilirubin decreased to normal levels. In a further paper [13] these workers reported on the effectiveness of tin protoporphyrin in suppressing excessive plasma bilirubin levels in anaemic mutant mice with profound haemolytic disease. Treatment resulted in substantial inhibition of haeme oxidase in liver, spleen and kidney and in a significant reduction of plasma bilirubin levels. The authors concluded that tin protoporphyrin has the ability to inhibit significantly in vivo, degradation of haeme and to diminish concurrently the plasma bilirubin levels in severe chronic haemolytic disorders.

Leishmaniasis is the name given to a group of infections common in tropical regions, caused by parasitic organisms of the genus Leishmania. These are flag-ellated protozoans of the family Trypanosomidae and the parasites, transmitted by small blood-sucking sand flies, are found in the skin or in deep organs (or both) of affected humans. The skin infections are marked by persistent granula-tomous and ulcerative lesions on exposed parts of the body. The mucocutaneous version of the disease affects the nose, mouth, pharynx and other mucocutaneous surfaces. Visceral leishmaniasis is an insidious chronic disease, resulting in a reduction in white blood cell levels, wasting and gross enlargement of the liver and spleen. None of the drugs used currently to treat the disease is de-void of significant toxicity and all must be administered in large doses over a long period of time. New, safer and cheaper leishmanicidal agents are badly needed, particularly those which can be given in a short intensive course or in a repository formulation of some kind. Trials have been described [14] with a number of experimental drugs in which di-n-octyltin maleate was amongst the most promising of the compounds tested. Further studies were proposed to examine the effectiveness of this organotin against cutaneous infections.

Complexes of organotins with Schiff bases have been tested as amoebicidal agents with promising results [15]. The most important amoebicides in current use are emetine and conessine and both are characterised by the presence of two nitrogen atoms in the molecule. It has been suggested that the drug-receptor association in these drugs is mediated by the formation of hydrogen bonds be-tween the nitrogen atoms of the drug molecule and specific bio receptors of the pathogenic organisms. In view of this, organotin complexes of sulphur-, nitro-gen- and fluorine-containing ligands were synthesised and examined for anti-microbial activity in vitro against the virulent strain of Entamoeba histolytica. Eleven Schiff base complexes were prepared and tested; the com-pound containing a tributyltin moiety:

had even greater activity than that of emetine. It was postulated that the activity was due to the presence of the tributyltin portion and also possibly to the presence of fluorine. In vivo studies were reported to be in progress and no claims have been made at this stage as to the viability of these organotins for practical therapy.

Another area in which organotin compounds play an important part is as intermediates in the synthesis of antibiotics. Some of these reactions have been reviewed by Smith [16]. In the preparation of antibiotics of the cephalosporin series, tributyltin esters of 7-aminocephalosporanic acids may be used to increase the solubility of these acids in a desired reaction medium. At the end of a reaction sequence involving the tributyltin ester, the free cephalosporanic acid may be re-precipitated by hydrolysing in the organic solvent with an excess of dilute acid. Tributyltin esters are also useful intermediates in the preparation of various penicillin derivatives. For example, 6-aminopenicillanic acid has been prepared by reacting the potassium salt with bis(tributyltin) oxide in benzene at 50-60°C under nitrogen, so as to produce the tributyltin ester. Treatment of the latter, preferably without isolation, with an imine-forming reagent (for example phosphorus trichloride), followed by subsequent hydrolysis under mild conditions, produces 6-aminopenicillanic acid. Bis(tributyltin) oxide has been used in the chemical synthesis of another antibiotic, the antibacterial agent nucleocidin. Sulphamoylation of the isolated 5'-hydroxyl group in 4'-fluoro-2',3'-O-isopropylidene adenosine is found to occur in high yield if it is carried out via the easily prepared 5'-O-tributylstannyl ether.

7.2 PREVENTIVE MEASURES BASED ON ORGANOTINS

In addition to the potential uses of organotins in therapeutic treatments, they may also find use in preventive measures aimed at eliminating the vectors of disease. Many illnesses which affect man, particularly in tropical environments, depend on parasitic organisms which are transmitted via intermediate hosts. Eradication of these carriers may then prove an effective means of controlling the spread of the parasite. There are several areas where organotins have shown potential for this type of approach and extensive trials have been conducted. The essential requirements are that the organotin biocide should show specific toxicity for the host without affecting other forms of life, should not introduce an environmental burden through its use and should be convenient and safe to handle.

7.2.1 Eradication of schistosomiasis

Schistosomiasis, or Bilharzia as it is commonly called, has not achieved the notoriety of smallpox or other world-scale diseases, but nevertheless, it is one

of the world's great scourges. The disease leads to chronic ill-health and de-
bility extending over many years and in addition to the suffering involved, it
leads to great loss in productivity and often to premature death of its victim.
It has been estimated [17] that between 200 and 300 million people are infected,
with an annual mortality of 2-4 million. Virtually unknown in temperate climates, the disease is widespread in Central and South America, Africa and Asia.

Fig. 7.3. Daily contact with infested water in tropical areas such as this poses
a constant health hazard from schistosomiasis.
Photograph courtesy: J. P. T. Boorman.

Schistosomiasis is the name given to a group of diseases caused by parasitic
blood flukes or schistosomes. These belong to the flatworm class and live within the blood vessels of their victim. There are three common forms of the disease: S. japonica is endemic to Asia and relates to liver and intestinal disorders, S. mansoni is found in Africa and Brazil and S. haematobium, known as
bladder bilharziasis, occurs in Africa, Cyprus and the Middle East. Schistosomiasis is transmitted to man when he contacts cercariae; the larvae are aquatic
and infectious for 8-12 hours. Human contact may be through drinking, bathing,
washing of clothes or domestic animals, swimming or merely traversing infested
water. The cercariae penetrate the dermis in 8-20 seconds, enter the blood-
stream and migrate to the capillaries in and around the liver and spleen. Here
they mature and as adult worms they reproduce sexually and lay a large number of
eggs, mainly in the exterior urinary tract and gut. The eggs move into the external environment during excretion and defaecation. On contact with non-saline

water, the eggs lie dormant and then eventually hatch into free-swimming mira-
cidial larval forms. These seek and enter specific species of snail (about 80
different species of snail are involved). Here, in the snail, the larva forms a
sporocyst. In this way, many cercariae of the same sex are formed by non-sexual
reproduction. When the sporocyst has matured, cercariae emerge and enter the
surrounding waters. They are then available to enter the human body and recom-
mence the cycle.

Medical problems arise as a consequence of the egg masses which remain in the
human body. These gradually block blood vessels and cause malfunctioning of vital organs. Cells in tissues die as they are slowly deprived of essential nu-
trients and oxygen. Outward signs of the disease include general weakness, impaired growth (in younger victims) and lessened resistance to infection; often,
mental faculties are impaired. Death is often due to lowered resistance to
pathogenic infections or to neoplastic growth within the bladder.

Possible methods of combating the disease include medication of affected individuals (not very effective), provision of clean water supplies, improved san-
itation and hygiene and elimination of the snail intermediate host. In practical terms, since many affected regions lack funds and skilled labour, the latter
approach is the most promising. Molluscicides have been used with some success
for many years in Japan and Egypt. The chemicals used were relatively simple:
copper sulphate, slaked lime and subsequently sodium pentachlorophenate. However these need skilled handling and repeated applications of large quantities.
For some years now, a vast number of chemicals have been screened with a view to
finding a molluscicide which would have a high and residual activity against
snails and their eggs, low toxicity to fish and to aquatic plant life and good
chemical stability.

In the 1960s many trialkyltin and triaryltin compounds were screened for molluscicidal activity and the data obtained have been reviewed by Duncan [18]. A
standard bio-assay procedure was established in which aquatic snails
Biomphalaria and Bulinus were exposed for 24 hours in an aqueous solution of the
candidate compound, followed by a 48 hour recovery period in water similar to
that used to prepare the molluscicide solution. Species differences for aquatic
snails were not usually very marked. The molluscicidal activity of tributyltin
and triphenyltin compounds was confirmed and field tests were begun throughout
the world, with various organotin compounds, particularly bis(tributyltin) oxide, T.B.T.O. Molluscicidal time-concentration relationships were worked out
for a number of organotin compounds to gauge their most effective application in
the field. Four compounds, tri-n-propyltin oxide, tri-n-butyltin acetate,

T.B.T.O. and butanestannonic acid were shown to be equal in activity to a proprietary organic compound, "Niclosamide" * over 24 hours exposure [19]. Further field tests, in Puerto Rico, showed the effectiveness of T.B.T.O. [20]. The site selected was a stream, part of which was cement-lined; this portion offered an opportunity for standardising the screening whilst the remainder offered a chance to study the impact of environmental factors such as vegetation. The snails, B. glabrata, were exposed in cages held at various distances from the point of application of the molluscicide. T.B.T.O. was applied at concentrations of 0.2, 0.5 and 1 ppm over 6-hour periods. The results were considered promising enough to suggest further testing.

Work by Deschiens et al. [21] in the Yaounda basin of the Cameroons, where intestinal bilharziasis is widespread, was aimed at determining the biocidal selectivity of T.B.T.O. The results showed that, whereas at 0.045 ppm there was a 70% kill of fish in the basin as well as a general destruction of snails, at 0.03 ppm all the snails were killed within 2-3 days and fish were unaffected. Even at 0.015 ppm T.B.T.O. there was a 100% kill of snails within 4-7 days. In further work, Deschiens [21] confirmed the high activity of T.B.T.O. in aquaria and in the field and also noted a residual activity against snails lasting 15 months (in aquaria) and 15 days (in the field). It was noted that low concentrations were effective against snails if the exposure period was prolonged.

This seemed to offer scope for the use of a slow-release system for dispensing the organotin molluscicide and an approach was developed, based on the use of impregnated rubber. (Antifouling applications of this type of material were discussed in Chapter 5.) The development of a suitable system and the results of practical trials have been described by Cardarelli [17]. Pellets of chloroprene rubber impregnated with the organotin can be floated on water inhabited by the snails, so as to release a small controlled dose of organotin toxic to the snails, over a long period of time (Fig. 7.4).

In preliminary trials, formulations releasing T.B.T.O., tributyltin acetate, or tributyltin sulphide were evaluated against Helisoma trivduis and excellent mortality rates were noted. Pellets retained their molluscicidal activity after daily washings over an 18-week period. In long term trials in the laboratory, several formulations remained active over a 26-month period. In further laboratory and field tests, 5 T.B.T.O. formulations and one other organic molluscicide were evaluated [22]. Pools and drainage ditches were used as test environments over a 4-month exposure period. Snail mortalities of 100% were achieved in the

* Trade name throughout.

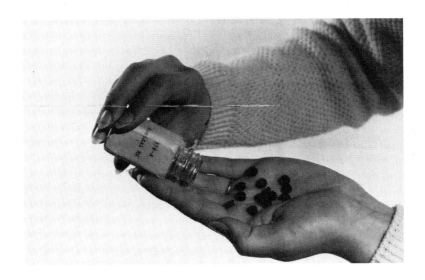

Fig. 7.4. A trial formulation for molluscicidal tests; rubber pellets containing T.B.T.O.

pool tests, and drainage ditch evaluations showed a continuous reduction in living snail populations even with flowing water conditions. In Tanzania, Fenwick [23] treated several pools with a formulation consisting of 8% "Bayluscide" * (a proprietary molluscicide) plus 3% T.B.T.O. in chloroprene rubber and observed a reduction in snail population over a 9-week period. It was observed that mortality of Biomphalaria pfeifferi varied with the dosage (72-96 hours at 4 pellets per litre and 192-216 hours at 1 pellet per 4 litres, for 100% mortality).

World Health Organisation approval of the use of T.B.T.O. for molluscicidal control is pending the result of long-term carcinogenicity studies which are in progress with this organotin, to ensure its environmental safety for this application. Advantages of organotin slow-release molluscicides have been summarised [24]. Since snail destruction is achieved through chronic intoxication, only ultra-low toxicant concentrations are necessary; moreover continuous low-level dosing of the water course means that repopulation through snail migration cannot satisfactorily occur. It is estimated that with the chloroprene-T.B.T.O. system, a transmission interruption of about 27 months could be achieved. The slow release organotins have also shown some potential for killing the cercariae released to water, but further trials are necessary.

7.2.2 Mosquito larvicides

The mosquito is a serious menace in many parts of the world, spreading malaria, yellow fever and other diseases. Controlled release organotins have been evaluated as potential mosquito larvicides [25]. It has been recognised for a number of years that trialkyltins and triaryltins are larvicidal. In studies, 6% T.B.T.O. in natural rubber, 20% tributyltin fluoride in natural rubber and 30% tributyltin fluoride in ethylene propylene copolymer, were evaluated, initially against larvae of Cx. pipiens and later, in standardised tests, against organophosphorus-susceptible, DDT-dieldrin-resistant Cx. quinquafasciatus larvae, pupae and adults. Neonatal laboratory mice served as a blood source for laboratory rearing. The criterion of effectiveness was the lethal time to cause 100% population mortality (LT_{100}). It had been observed earlier that the organotins dramatically slow or prevent morphogenesis. The normal egg-to-adult time of 8-9 days under these test conditions was extended to 21 or more days at sublethal doses. Pellets of biocidal rubber were soaked continuously prior to, and during the test periods. The data were taken to indicate long-term efficacy for several controlled release organotins against culicine larvae under laboratory conditions. Long-term effectiveness, low biological persistence (following release) and the suspected proteolytic kill mechanism which may negate the development of tolerance, are all considered to indicate that T.B.T.O. and tributyltin fluoride in controlled release formulations are suitable as mosquito larvicides.

Cardarelli [26] has described slow release systems based on monolithic incorporation of the toxicant in thermoplastic polymers containing a porosigen which gradually solvates into an external water medium, leaving a porous network through which the toxicant may be gradually leached. A typical system consists of an ethylene vinylacetate copolymer 28.6 parts, low density polyethylene 27.6 parts, zinc stearate 2.4 parts, calcium carbonate 17.0 parts and tributyltin fluoride 30 parts. Formulations of this type have been tested as molluscicides and insect larvicides. Long-term effectiveness has been demonstrated against 1st and 2nd instar larvae of C. pipiens quinquafasciatus in short-term and long-term tests.

REFERENCES

1 A.J. Crowe, P.J. Smith and G. Atassi, Chem. Biol. Interact., 32 (1980) 171-178.
2 Screening data summary interpretation, U.S. Nat. Cancer Inst. Instruction 14, 1978.
3 A.J. Crowe, International Tin Research Institute (Private communication).
4 P.J. Sadler, Chem. Britain, 18 (1982) 182, 184, 188.
5 E.J. Bulten and H.A. Budding, Brit. Pat. 2,077,266A, 1981.
6 S. Ozaki and H. Saki, Jap. Pat. 76-100,089, 1975.
7 R. Barbieri et al., Inorg. Chim. Acta, 66 (1982) L39-L40.
8 I. Haiduc, C. Silvestru and M. Gielen, Bull. Soc. Chim. Belg., 92 (1983) 187-189.

9   W. Seinen, Vet. Sci. Commun., 3 (1979/1980) 279-287.

10 A. Kappas and M.D. Maines, Science, 192 (1976) 60-62.

11 T. Yoshinaga, S. Sassa and A. Kappas, J. Biol. Chem., 257 (1982) 7778-7785.

12 G.S. Drummond and A. Kappas, Science, 217 (1982) 1250-1252.

13 S. Sassa, G.S. Drummond, S.E. Bernstein and A. Kappas, Blood, 61 (1983) 1011-1013.

14 W. Peters, E.R. Trotter and B.L. Robinson, Ann. Trop. Med. Parasitology, 74 (1980) 321-335.

15 A.K. Saxema et al., J. Toxicol. Environ. Health, 10 (1982) 709-715.

16 P.J. Smith, Chem. Ind. 4th Dec. 1976, 1025-1029.

17 N.F. Cardarelli, Controlled Release Pesticidal Formulations, C.R.C. Press, Cleveland, OH, U.S.A., 1976.

18 J. Duncan, Pharmac. Therapy, 10 (1980) 407-429.

19 L.S. Ritchie, L.A. Berrios-Duran, L.P. Frick and I. Fox, Bull. W.H.O., 31 (1964) 147-149.

20 L.A. Berrios-Duran, L.S. Ritchie and H.B. Wessel, Bull. W.H.O., 39 (1968) 316-320.

21 R. Deschiens, H. Brottes and L. Mvogo, Bull. Soc. Pathol. Exotique, 59 (1966) 968-973.

22 C.P. Da Souza and E. Paulini, W.H.O. Rep. Ser. PD/MOL/69.9, 1969.

23 A. Fenwick, W.H.O. Rep. Ser. AER/BILHARZ/14 Add. 1, May 21st, 1969.

24 S.D. Hagen, S.V. Kanakkanatt and N.F. Cardarelli, Proc. Inf. Conf. Schisto., 1978, pp. 459-467.

25 N.F. Cardarelli, Mosquito News, 38 (1978) 328-333.

26 N.F. Cardarelli, 8th Internat. Symp. Controlled Rel. Bioactive Materials, Ft. Lauderdale, FL, U.S.A., 26th-29th July, 1981, pp. 117-132.

Chapter 8

MISCELLANEOUS BIOCIDAL USES

8.1   INTRODUCTION

In addition to the applications of organotins already described there are a
number of other uses where their biocidal properties are employed to advantage.
These include the field of materials protection, where they have been used to
protect stonework, paints, textiles, leather and cement. Organotins have been
used as anthelminthics in animal husbandry and as disinfectants in hospitals and
on clothing. In all these applications, the potential of organotins for attacking damaging organisms without being toxic to higher life forms and for eventually degrading to harmless inorganic forms of tin is an advantage.

8.2   MATERIALS PROTECTION

8.2.1   Protecting masonry and stonework

Stone and masonry structures are susceptible to attachment of biological
growths under certain conditions and whilst in many cases this type of weathering is encouraged as being attractive in appearance, there are other instances
where such growths are not welcomed, for example on headstones of graves or on
concrete or stone pathways, where they impart a dirty or neglected aspect. In
addition, mosses and lichens may secrete acids which can cause etching and often
significant deterioration of carbonaceous stone. Heavy accumulations of growth
will keep stone surfaces permanently damp, increasing the risk of damage through
freezing or corrosion. The problem of controlling such biological growths has
been reviewed by Richardson [1]. There are several types of micro-organism involved in the weathering of stone or masonry. Algae will develop on porous natural stone whenever a suitable combination of dampness, warmth and light occurs.
These algae may be bright green in colour and occasionally dark green, brown or
pink. During dry weather these die off, leaving a limited dirty deposit of dead
algae, giving an unsightly appearance; the spores can regenerate when suitable
conditions return. Mosses grow less rapidly and usually require the presence of
organic deposits already on the stone; dead algae can supply this requirement so
that algal growth followed by mosses can be considered as a process of colonisation taking place on the stone. Lichens are fungi, usually Ascomycetes or Basidiomycetes, with algae normally present inside the structure in symbiotic association and determining the colouration of the growth. Lichens grow very
slowly, the diameter of the thallus increasing typically at a rate of 1 mm per
annum in north-west Europe. Beneath this visible thallus, hyphae penetrate the

stone to an appreciable depth. Lichens can withstand considerable extremes of temperature and humidity.

Monumental features in porous limestone and sandstone are particularly susceptible to growth, but problems are also frequently encountered on buildings, especially in areas of high rainfall. Mosses, lichen and algae may be removed mechanically by scraping or wire brushing followed by scrubbing with clean water. This will leave a stone surface adequately clean with the exception of rings which mark the extremities of lichen thalli. However, algae develop from airborne spores and growth will recur whenever suitable conditions exist; in fact the washing process may even encourage the subsequent rapid development of algal growth. Lichen hyphae remaining within the stone can also cause a fresh growth in a relatively short period. External water repellency treatments may be designed so as to control these growths on porous masonry partially, but incrustations already in place must be removed first. The conventional agents used to remove such growths, although effective, can have certain disadvantages, causing staining of light-coloured surfaces and possible damage to the stone surface by chemical attack.

A relatively new treatment which has proved effective against persistent growths, giving long-term protection, is based on T.B.T.O. used in combination with a quaternary ammonium compound. The quaternary solubilises the organotin and also has a biological activity of its own, rupturing the cell walls of algae and lichens as well as exerting a toxic action. The powerful fungicidal properties of T.B.T.O. have already been discussed in Chapter 4. The effectiveness of these organotin-based treatments was established during extensive trials conducted by the Penarth Research Centre in Winchester, U.K. for the Commonwealth War Graves Commission in 1969. Headstones and other monuments in the care of this Commission must never appear to be neglected and must present a clean, bright appearance. Fig. 8.1 shows the type of improvement which has been obtained using a proprietary formulation based on T.B.T.O. and a quaternary ammonium compound. Several proprietary formulations of this type are now available commercially.

More recently, tests have been conducted at the U.K. Building Research Centre into masonry paints and cleaning methods for walls affected by organic growth [2]. Eight biocidal treatments (including a T.B.T.O./quaternary formulation) were used in combination with various cleaning procedures and bands of commercial paints were applied. The results threw greatest emphasis for protection on the mode of cleaning. Water jet cleaning was suggested as the minimum, with abrasive cleaning for best possible results.

234

Fig. 8.1. These headstones provide a graphic example of the effectiveness of
Thaltox Q * in removing moss and algae.
Photograph courtesy: Wykamol Ltd., Winchester, Hants.

8.2.2 Paint additives

One area where triorganotin biocides have potential, but have not been widely
used to date, is as preservatives for paints. There are two problems involved
in protecting paints; firstly there is the need for in-can protection and secondly there is the need to protect paint films after application to surfaces.
Triorganotin compounds are not particularly effective as in-can preservatives,
but have proved useful in protecting coatings of both water-based and oil-based
paints against attack by micro-organisms, especially in damp rooms and under un-
favourable climatic conditions. Tributyltin compounds most suitable for this
application include the oxide, fluoride, acetate, benzoate, naphthenate, linoleate and abietate. These are all compatible with most binders in the quantities
required for effective biocidal activity and do not cause yellowing. Tributyltin compounds can be used in aqueous emulsion paints (polyvinyl acetate, sty-
rene-acrylic, acrylic and epoxy resin emulsions) and in solvent-based paints.
Tests must first be conducted with aqueous emulsion paints to ensure that the
normal amount of emulsifying agent is sufficient to disperse the water-insoluble
organotin, and more agent may be needed on occasion. Proprietary triorganotin
condensates ("polyoxilates") have been produced which will disperse in water in
any ratio. They are claimed to be insensitive to fluctuations in pH, electro-
lytes, and weak or dilute acids or alkalis and are marketed under the Trade name

* Trade name throughout

"Metatins" by Acima AG of Switzerland. Liquid tributyltin compounds can be simply stirred into the paints. Solid organotins such as tributyltin fluoride must be dispersed as finely as possible to obtain homogeneous paints and grinding the organotin with the paint during manufacture is recommended. Tables 8.1 and 8.2 present data which have been acquired in laboratory studies [3].

TABLE 8.1

Effect of organotins on PVA emulsion paint films [3]

Biocide		Aspergillus niger		Bacillus subtilis
		Growth on sample	Inhibition zone mm	Inhibition zone mm
T.B.T.O.	0.5%	–	1-2	5-6
	1.0%	–	2-3	8-9
	2.0%	–	4-6	10-15
Tributyltin fluoride	0.5%	–	1	5-6
	1.0%	–	2	7-8
	2.0%	–	4	10-12
No treatment		+++	0	0

TABLE 8.2

Effect of organotins on alkyd resin paint films [3]

Biocide		Aspergillus niger		Bacillus subtilis
		Growth on sample	Inhibition zone mm	Inhibition zone mm
T.B.T.O.	0.5%	–	0-1	4
	1.0%	–	2	5
	2.0%	–	3-4	8-9
Tributyltin fluoride	0.5%	–	0	2
	1.0%	–	1-2	3-4
	2.0%	–	3	6-8
Tributyltin linoleate	1.0%	–	0	1-2
	2.0%	–	1-2	3-4
	3.0%	–	3	6-7
No treatment		++	0	0

A comprehensive review of the use of organotin biocides in paint systems was published by Bennet and Zedler [4]. Drisko [5] tested 15 paints incorporating five resin variations and three pigment variations for mildew resistance. A high acid number alkyd resin which gives a higher tin content when reacted with T.B.T.O. was used for most of the paint systems. In some formulations the organotin additive consisted of the reaction product of T.B.T.O. with 3, 4, 5, 6

tetrachlorophthalic anhydride which also has some biocidal activity. Zinc oxide
was included in some formulations to see whether it had any synergistic action
with the organotin compound. In laboratory studies over 8 weeks, no growth occurred on paints containing zinc oxide and organotin. Panels coated with paints
containing barium metaborate or the reaction product of T.B.T.O. with the tetrachlorophthalic anhydride as the sole active ingredients showed significant
growth after 8 weeks.

8.2.3   Textile protection

Organotin biocides are used to protect certain textile fabrics against insect
attack and also against micro-organisms causing rotting of fabric. Textiles
such as wool are subject to extensive damage by moths and carpet beetles. One
paper [6] has drawn attention to the possibilities of triphenyltin compounds for
permanent mothproofing of wool. Triphenyltin acetate and triphenyltin chloride
are colourless crystalline solids, insoluble in water but sufficiently soluble
in common organic solvents for emulsifiable concentrates to be formed. They are
not steam-volatile and may be exhausted on woollen fabrics from a boiling aqueous emulsion within 30 minutes at pH 3-7. There are indications that the triphenyltin group may be in part covalently bound to the wool fibre. A calculated
application of 0.025% of the triphenyltin compound was shown to protect wool
from the ravages of the clothes moth and the carpet beetle and to maintain its
action even after 20 laundering cycles in soap solution or 20 cycles of dry
cleaning in white spirit. Typical dye-bath liquors and rinsings contained 0.1-
0.3 ppm of organotin residues, which corresponds to exhaustion values exceeding
99% after treatment.

In subsequent work [7] the insect-proofing properties of triphenyltin chloride were examined in greater detail. An emulsifiable triphenyltin chloride
concentrate was added to a dyebath in which a soap-scoured, undyed, plain-weave
woollen fabric was steeped and optimum conditions established for the rate of
exhaustion of triphenyltin chloride. Good levelling properties were found and
there was no deleterious effect on the dyeing behaviour with the dyes in use.
The minimum effective concentration for insect proofing was established as
0.05%. At this level, the organotin was a larvicide for the clothes moth and a
pronounced antifeedant for the carpet beetle during the 14-day bio-assay (see
Table 8.3).

Marginally satisfactory mothproofing was obtained when moth-proofed wool was
subsequently given a reductive or oxidative bleach, or when the organotin was
applied during an oxidative bleach or after reductive bleaching. The treatment
was still effective after 20 hand-laundering cycles or 20 dry-cleanings or after

TABLE 8.3

Minimum effective concentration of triphenyltin chloride against wool-attacking insects [7]

Insect	Feeding damage (mg) at indicated concentrations * %				
	0.01	0.05	0.1	0.3	0.5
Clothes moth (Tineola)	22	4	4	2	2
Carpet beetle (Anthrenus)	42	6	5	3	3

* Wool is considered "moth-proof" when feeding damage does not exceed 8 mg (provided control values exceed 35 mg).

a 90-day exposure to sunlight. (This apparent photochemical stability once the organotin has been absorbed by the wool, contrasts with the rapid degradation which has been found in agricultural applications of triphenyltins.)

Analytical determination of residues associated with the use of triphenyltin chloride indicated that exhaustion exceeded 99.7% when the organotin was applied in a hot dye-bath. The associated residues were readily decomposed by persulphate, indicating that it should be possible to moth-proof wool without risk of contaminating the aquatic environment with waste waters. Mill trials in which up to 900 kg of wool were used, confirmed these conclusions. Studies on leaching effects of perspiration and saliva suggested that it was unlikely that a hazardous amount of the organotin would be extracted, but it was recommended that baby wools should not be treated.

Textiles based on natural materials such as cotton, silk and wool are also prone to attack by micro-organisms with a resultant loss of strength and discolouration. Cellulose-based products (cotton or jute) can serve as a food source for many micro-organisms, of which Chaetomium globosum is typical. Although some of the damage can be prevented by taking suitable measures to avoid high humidity or soil contact, in many cases a biocidal treatment of the textile is necessary. Amongst the many biocides used for textiles, T.B.T.O. and tributyltin linoleate or naphthenate are effective against a wide range of fungi and bacteria. The organotins may be applied to textiles from organic solvents or from aqueous emulsions. They are not easily washed out of the fabric, since they have a high affinity for cellulose. The trialkyltin compounds are also normally compatible with other biocides and with the various commonly used textile auxiliaries. A frequently used test method is the DIN 53931 growth test in which textile samples are exposed on a nutrient agar medium to micro-organism growth, and the presence or absence of growth noted and the inhibition zone measured. T.B.T.O. and tributyltin linoleate have been shown to impart distinct biocidal activity to treated cotton fabrics against <u>Chaetomium globosum</u> (see Table 8.4).

TABLE 8.4

Effect of tributyltin compounds on growth of <u>Chaetomium globosum</u> on cotton

Biocide		Growth			
		Without leaching		After leaching in running water	
		Growth	Inhibition zone	Growth	Inhibition zone
T.B.T.O.	0.1%	–	2–3	–	2
	0.2%	–	5–6	–	4–5
	0.5%	–	10	–	8–9
Tributyltin linoleate	0.1%	–	1–2	–	1–2
	0.2%	–	3	–	2–3
	0.5%	–	6–7	–	6
No treatment		+++	0	+++	0

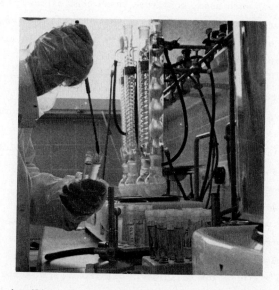

Fig. 8.2. Sterile handling of micro-organisms in the laboratory
Photograph courtesy: British Sanitized Ltd., Ashby-de-la-Zouch, U.K.

A more rigorous test is the soil burial test in which textile samples are buried and exposed to the action of natural soil micro-organisms for four weeks. Loss of strength is then measured and used as an assessment of biological attack. In such tests twilled cotton fabric, treated with 0.1% or 0.5% T.B.T.O. has shown little loss of strength after testing. The use of actinomycetes in testing biocides for textile protection has been described [8]. Many actinomycetes can degrade cellulose and deterioration of untreated cotton by this organism has often been reported. Starched and unstarched pure cotton standard test cloth was impregnated with various preservatives, including 0.1% and 0.2% T.B.T.O. and placed on the surface of cellulose agar; it was dampened with distilled water and Streptomyces spores inoculated on to the textile. Growth was visually assessed after incubation at 30°C for 5 days and the tensile strength measured. At the 0.2% level, T.B.T.O. preserved the tensile strength of the test fabrics but did not prevent pigment production and staining of the fabric. Carraher et al. [9] described chemical modification of cotton to enhance its thermal and biological properties. Cotton was dissolved in bis ethylenediamine-copper(II) hydroxide and reacted with organic solvent containing dibutyltin dichloride or dipropyltin dichloride. Cotton in normal or modified form was precipitated by neutralising the solution with dilute sulphuric acid. The modified product was considered to consist of a mixture of mono-, di- and tri-substituted material with the relative amounts of di- and tri-substitution increasing as the proportion of the tin moiety increased. The solid products were generally flexible and in biological testing on paper discs carrying spores of typical organisms which attack cotton, they exhibited good inhibition of fungal growth.

An interesting potential application of T.B.T.O. is in synthetic fibre/cotton blends for mine belting [10]. Prior to the Cresswell Colliery fire disaster in 1950, conveyor belts were almost completely constructed of rubber/cotton blends, but this was gradually changed to PVC/cotton. Problems of biodeterioration became more pronounced with this blend, since damage to the PVC covering, during use, exposes the cotton to micro-organisms arising from infected water or from ventilation systems. T.B.T.O. at a concentration of 0.2-0.5% was shown to give protection to the cotton in this application. The preservative can be introduced either by pad application to the textile core of a ply belting prior to production or by incorporation of the fungicide in the plastisol and subsequent transfer to the textile core during curing. Table 8.5 shows results obtained with orthophenyl phenol in simulated mine belting.

8.2.4   Protection of other materials

Although plastics are not particularly vulnerable to attack by micro-organisms, under suitable conditions their quality can be impaired. In most cases,

TABLE 8.5

Preservative efficiency of T.B.T.O. and o-pp in simulated mine belting

Fungicide		Residual tensile strength of textile core after soil burial, %		
		1 month	2 months	3 months
None		39	2	0
Orthophenyl phenol (o-pp)	3%	77	35	18
T.B.T.O.	0.25%	100	60	20
	0.5%	100	80	45

it is not the polymer itself which serves as food for the micro-organisms, but rather the processing aids, such as plasticisers or lubricants. Tributyltin compounds can give long-lasting protection when incorporated into polymer blends, the degree of activity being relatively independent of the type of plastic [3]. Polyvinyl chloride, unsaturated polyesters, epoxy resins and polyethylene are amongst the plastics which can be protected in this way. PVC sheeting has a wide range of end uses, as a roofing material, swimming pool liners and vinyl flooring. Introduction of an organotin biocide at the processing stage ensures that the whole of the polymer is uniformly protected. It then prevents microbial attack of the plasticiser and the occurrence of staining due to micro-organisms producing pigments which are very easily absorbed by the PVC. Polyurethanes are unusual amongst man-made polymers in that they are directly susceptible to fungal attack. When fabrics are backed with polyurethane foam, as in lightweight rainwear, the garments are frequently rolled up after use and may be stored whilst wet, ideal conditions for mildew and mould to develop quickly. Any attack of the actual polymer can lead to small cracks appearing in the backing and hence to loss in waterproofing ability. A suitable organotin system can be incorporated during manufacture of the polyurethane. In this case, where one is working with a highly reactive system, great care has to be exercised in choosing a biocide which will not influence other essential properties (for example, resistance to hydrolysis), nor react with the system so as to inactivate the biocide.

Sealing materials such as silicones, polysulphides, polyacrylics, polyurethanes and bitumen-based materials can also be protected against microbiological attack by incorporation of up to 1% of a tributyltin compound such as T.B.T.O. An organotin-based formulation is also available commercially for protecting old books, manuscripts, etchings, etc., from insecticidal attack. Books and their bindings are susceptible to damage by four or five species of insect and several varieties of fungal mould, when stored in a damp environment. The formulation "Bibliotex" *, marketed by Wykamol Ltd., of Winchester, U.K., is based on T.B.T.O. and hexachloro-cyclohexane and is applied by spraying lightly the affected pages. Synthetic chamois cloths are made by coating a cotton fabric with an acrylic resin. The item is usually stored wet or damp and mildew can easily develop, rotting the cloth. Use of 0.3-0.8% of an organotin-based treatment incorporated in the acrylic coating (and possibly also applied to the cotton fabric) will prevent degradation, mildew and unpleasant odours.

8.3 SLIME PREVENTION IN THE PAPER INDUSTRY

The paper industry recirculates industrial water to reduce wastage and this favours the occurrence of slime forming bacteria and fungi. Slime is a viscous or hard mass, consisting of a mixture of micro-organisms and pulp and its accumulation may interfere with paper production, by clogging equipment; moreover paper quality may be impaired. It is difficult to destroy or prevent the slimes since a wide range of micro-organisms can give rise to them. Organomercury compounds and pentachlorophenates have been used as a control measure, but their toxicity is a limitation. Tributyltin and tripropyltin compounds can give effective protection against slime formation. It has been claimed that as little as 0.06 ppm of T.B.T.O. is sufficient to lower the concentration of bacteria and moulds in the water to a level where they do not interfere with paper production [11]. Emulsifying agents are required to ensure effective dispersion of the tin compound in water. Combinations of T.B.T.O. or tributyltin benzoate with quaternary ammonium compounds are effective slimicides, counteracting not only bacteria but also numerous fungi and algae [12]. Thust [13] has described how studies in the German Democratic Republic have shown that organotin compounds are adsorbed very quickly and to a high degree on the mass of cellulose and waste paper, so that their use does not impose an environmental hazard. A formulation containing organotin compounds, chlorinated phenols and quaternary ammonium compounds for preventing build-up of organisms in underground gas reservoirs has also been described by this author.

8.4 ORGANOTIN-BASED DISINFECTANTS

When considering the potential of trialkyltin compounds for disinfection, it is necessary to distinguish between bactericides and bacteriostats. A bactericide must kill all micro-organisms present within a specified time, a bacteriostat operates by inhibiting the possibilities for growth of the bacteria and in the long term leads to a reduction in germs due to physiological withering of the organism. Organotins are essentially bacteriostatic in action; furthermore, whilst even at very low concentrations they are effective against gram-positive bacteria, they are less so against gram-negative ones. The organotins are therefore best used in combination with a fast-acting bactericidal agent and this is the approach adopted in commercial formulations. Formaldehyde is a convenient bactericide to use with the organotin and combinations of tributyltin benzoate and formalin have proved to be excellent disinfectants with long-term activity. Nösler [14] has described the development of a commercial product of this type, "Incidin" *, marketed by Henkel & Cie GmbH, Düsseldorf. Testing against Staphylococcus aureus on PVC surfaces showed that tributyltin benzoate has long term biocidal action and that this biocidal activity increased with successive treatments of the surface; thus a permanent disinfectant effect can be achieved, since the organotin compound is strongly held on the surface. This was confirmed in disinfection studies at a hospital, using test plates on which germ colonies were estimated. To avoid carry-over of the bacteriostat to the test plate, which could give rise to incorrect results, an inactivator such as Tween 80 * was applied to the test plate before counting; this neutralises any biocide present. It was shown that germ levels were considerably reduced by treatment with Incidin. The two biocides exert a synergistic effect on each other, since organisms which have been damaged by the organotin compound, but not killed, are destroyed by smaller quantities of formaldehyde than when the organotin compound is not present. One commercial product, "Kombinal asept" *, which is marketed by VEB Chemiekombinat Bitterfeld in the German Democratic Republic, contains 3% tributyltin benzoate, 10% formaldehyde and 3% of a quaternary ammonium salt.

Results of a number of hospital disinfection trials have been reported in the literature. As early as 1959, Hudson et al. [15] described trials in a hospital in the U.S.A., using an agent based on T.B.T.O. plus a quaternary ammonium compound. The trials were aimed at reducing the risk of cross-infection due to Staphylococci. This was incorporated into paint for the hospital walls, waxes and mopping fluid for the floors, and rinses for the laundry, in an attempt to make the hospital self-decontaminating. The biocides were also introduced into the standard hospital air filters. In areas where, before treatment, cultures showed staphylococcus counts of 40-50 colonies, after the introduction of the treatment, counts of 5-10 colonies were found and maintained at these lower levels over 8 weeks. Rees [16] described similar tests in a maternity home in Cape Town. Nursery and hospital sluice rooms were decontaminated using T.B.T.O.-containing paint for the walls, wax for the floors and rinses for the nursery laundry. The floor of each room was mopped daily with T.B.T.O. solution. Results appeared to corroborate those found in earlier studies.

However, some conflicting results have been reported. In one study [17] a number of different techniques for disinfecting and cleaning hospital ward floors were tested, using a series of 14 disinfectants, including a tributyltin compound combined with a quaternary ammonium compound. Laboratory tests, surface disinfection trials and ward studies were conducted. Impression plate samples showed little or no reduction in total bacteria or in Staph. aureus on exposed floors after washing or disinfection; when an area of floor was protected from re-contamination by inverting an open box over the area, large reductions (93-99%) in total bacterial counts were observed. The authors concluded that if floors were to be kept bacteriologically clean, it would be necessary to reduce the access of bacteria by air and by contact; air-ventilation ducts had not been treated in this study.

Schulz-Utermöhl [18] studied combinations of organotins with formaldehyde and cleaning agents, to discover whether the latter had an inhibiting effect on the biocidal activity, since certain compounds can inactivate the biocides. In fact, an improved, synergistic effect was found with the agents tested. One paper [19] describes how treatment of locker and training rooms of a U.S. football team with an organotin biocide, reduced the number of staphylococcal foot infections suffered by football players during an abnormally dusty season. Verdicchio [20] evaluated tributyltin benzoate plus a quaternary ammonium compound as a hard surface disinfectant for hospital floors. Following successful in-vitro tests and estimates of residual activity in a high-traffic corridor, evaluations were carried out at a Florida hospital. Synergistic effects were observed between the organotin and the quaternary against Pseudomonas aeruginosa as test organism. Enhanced bactericidal activity was also demonstrated in-vitro against myobacterium tuberculosis. It has been reported [21] that a formulation based on T.B.T.O. and a quaternary ammonium compound (Tylon A-35 *, manufactured by Vestal Laboratories Division of Chemet Corp., Missouri, U.S.A.) has shown promising results in preventing the growth of Legionella pneumophila in cooling towers. It is believed that a prime cause of outbreaks of "Legionnaire's Disease" is the spread of this pathogenic bacterium via the cooling towers or evaporative conductors of air conditioning systems. The bacteria may be spread via the circulated air after being picked up from the water in the cooling tower. It has been suggested that regular use of Tylon A-35 in cooling towers, combined with an effective cleaning and maintenance programme, would contribute greatly to eradicating the possibility of further outbreaks of "Legionnaire's Disease".

Textile floor coverings now hold a substantial sector of the overall flooring market, not only in homes but also in the public sector, in hospitals, clinics,

244

Fig. 8.3. Non-woven carpet material treated with antimicrobial organotin compounds in a rat test for establishing safe effective levels.
Photograph courtesy: British Sanitized Ltd., Ashby-de-la-Zouch, U.K.

hotels, schools and institutions. Such communal areas can lead to an increase in the build-up of germs such as the athlete's foot fungi which lie dormant for long periods before being picked up and passed on to another person. Organotins have been used in commercial antimicrobial treatments which reduce germ counts in textile floor coverings and in a variety of other materials. A range of treatments, some embodying organotins, are manufactured by British Sanitized Ltd., under the trade names "Actifresh" * or "Sanitized" *. These can be applied to textile floor coverings during the wet-finishing process of the textile fibres and/or it can be included in backing materials. This is particularly relevant for fibre-bonded type carpets where acrylic resins are extensively used and it may be much easier to incorporate the Sanitized treatment at this stage rather than treating the fibres separately. The treatment can also be incorporated into a silicone or fluorocarbon finish which is applied together or in conjunction with, an acrylic coating for roller blinds or shower curtains. These will prevent development of micro-organisms under the conditions of high humidity found in kitchens and bathrooms.

Such treatments are also applied to protective clothing, where perspiration tends to collect on the inside of the garment, be decomposed and produce offensive odours. Use of a suitable treatment can improve hygiene, retard odour development and preserve the aesthetic and physical qualities of the textiles.

8.5 ORGANOTINS IN ANIMAL HUSBANDRY

Certain organotin compounds have found a number of applications in animal husbandry as anthelminthic agents in poultry, and as insecticides for sheep and cattle. Absence of disease is one of the leading factors in raising poultry successfully, alongside correct breeding and good nutrition. The early poult stage of development is one at which birds are particularly vulnerable to a number of parasitic infections. Worm infestations, although they cause a relatively minor death toll, are responsible for retarded growth, loss of meat and egg production and wasted feed and labour.

Three types of worm are involved: large roundworms which spread themselves by the droppings of birds and which penetrate the lining of the intestine, cecal worms which inhibit the blind ends of the ceca, and tapeworms which are transmitted as eggs via an intermediate host (earthworm, slug, snail) which is eaten by the bird. Di-n-butyltin dilaurate has been shown in experimental studies [19] to be capable of removing a wide range of species of tapeworm which affect poultry. Results are summarised in Table 8.6.

TABLE 8.6

Tapeworms against which dibutyltin dilaurate has shown activity [19]

Tapeworm	Effectiveness %	Location	Intermediate host	Affected poultry
Davainea proglottina	86-100	Duodenum	Slugs, snails	Chicken
Amoebotaenia sphenoides	85-100	Duodenum	Earthworms	Chicken, turkey
Hymenolepis carioca	85-100	Duodenum	Stable flies, dung beetles	Chicken, turkey
Raillietina cesticillus	85-100	Jejunum	Houseflies, beetles	Chicken, turkey
Choanotaenia infundilailum	65-100	Jejunum	Houseflies, beetles, grasshoppers	Chicken, turkey
Raillietina tetragono	91-100	Ileum	Ants	Chicken, turkey

"Wormal" * tablets or granules, manufactured by Salsbury Laboratories of Charles City, Iowa, U.S.A., contain a mixture of dibutyltin dilaurate, piperazine (to combat roundworms) and phenothiazine (active against cecal worms). When one pound of Wormal granules are mixed into 100 pounds of feed, the mixture will treat 250 chickens or turkeys until consumed so that each bird gets a recommended dosage. Worming schedules have been worked out for chickens (broilers or replacement stock) and turkeys. One paper [20] has however described some yolk mottling occurring in eggs laid by chickens which had been treated with dibutyltin dilaurate, above or in combination with piperazine and phenothiazine. The effect was reported to be most apparent 7 days after treatment, after which it rapidly decreased; egg weight and shell thickness were not affected.

Coccidiosis is a disease caused by one or more species of coccidia, microscopic single-celled parasites of the protozoan division of animal life. The disease normally occurs in poults 2 to 10 weeks old and is spread from bird to bird through the eating or drinking of contaminated food, water or litter. Coccidiosis can be a problem in turkeys reared in confinement, leading to poor feed utilisation and stunted growth. "Tinostat" * tablets (from Salsbury Laboratories) containing dibutyltin dilaurate as active ingredient are claimed to be effective in preventing development of this disease when fed at a rate of 3 pounds per ton of feed, over the first 10 to 12 weeks of the turkey poult's life.

Fig. 8.4. Commercial anthelminthic agents based on dibutyltin dilaurate. Courtesy: Salsbury Laboratories, Charles City, Iowa, U.S.A.

Workers have also reported anthelminthic activity in diphenyltin dichloride and dioctyltin dichloride when tested with chickens [21]. The recommended dose was 250-300 mg/kg of feed over a 20-hour diet. However, toxic effects were manifested in some cases at these dosages.

Deufel [22] has described the effectiveness of dibutyltin oxide in controlling worm infestations in rainbow trout. A dose of 250 mg/kg of fish weight, incorporated in the feed, was shown to be effective against Acanthocephala in the fish. A number of trimethyltin compounds have been evaluated as potential agents for preventing attack by sheep blowfly (Lucilia sp.) [23]. The economic loss attributed to sheep blowfly in the world's major wool-producing countries amounts to many millions of dollars each year. Insecticides have long been used for prophylaxis and treatment of fly strike, but strains resistant to the chlorinated hydrocarbon insecticides have made their appearance. Trimethyltin compounds and other insecticides, have been screened for dermal toxicity to mice, persistence on shorn wool, dermal toxicity for sheep and persistence on wool after jet spraying. Compounds selected for more extensive evaluation included trimethyltin chloride at 0.025% and 0.05% concentration; this persisted on shorn wool for the entire time of trial (24 weeks) and was well tolerated by sheep at dosages 2-3 times in excess of those needed for effective treatment. Persistence on shorn wool of 20-25 weeks and on wool on the live animal of more than 13 weeks was taken to suggest the feasibility of using trimethyltin chloride to prevent blowfly strike by application before an anticipated fly wave. The authors, however, point out that care must be exercised in testing the trimethyltin compounds further, until their safety to humans during application has been definitively established.

Tricyclohexyltin hydroxide has been tested, along with a number of other acaricides, against the sheep scabies mite Psoroptes ovis [24]. Common scabies of sheep, also called "sheep scab", is a highly contagious and debilitating disease. The scabies mite can also cause scabies in cattle and horses and in fact is at present only found in cattle in the U.S.A., sheep scab having been eliminated there since 1937. However, it is found in sheep in the U.K. and in other parts of Europe. Scabies-infested sheep from a closely-quarantined test flock in the U.S.A., were treated by dipping in baths of the various candidate acaricides and then isolated for the period of the test. Trials were considered a success if no live mites and no irritation were present 2-3 months after treatment. Tricyclohexyltin hydroxide was considered marginally effective in the range 0.1-0.125% and trials at somewhat higher concentrations were suggested.

REFERENCES

1   B.A. Richardson, Stone Ind., 8 (1973) 2-6.
2   P. Whitely and A.F. Bravery, J. Oil Col. Chem. Assn., 65 (1982) 22-27.
3   Organotin compounds: Biocidal protection of various materials, Schering AG
    Publication, 1983, 32 pp.
4   R.F. Bennet and R.J. Zedler, J. Oil Col. Chem. Assn., 49 (1966) 928-953.
5   R.W. Drisko, L.K. Schwab and T.B. O'Neill, Proc. Controlled Release Pest.
    Symp., Oregon State Univ., 22nd-24th Aug. 1977, 181-186.
6   R.M. Hoskinson and I.M. Russell, Austr. J. Text. Inst., 64 (1973) 550-551.
7   R.M. Hoskinson and I.M. Russell, Austr. J. Text. Inst., 65 (1974) 455-463.
8   T.R. Fermor and H.O.W. Eggins, Internat. Biodet. J., 17 (1981) 15-17.
9   C.E. Carraher et al., Organic Coatings Plast. Chem., 40 (1979) 560-565.
10 Albright and Wilson Ltd., Tech. Service Note DEV-TS/JLB/BJH, 31st Jan. 1972.
11 German Pat. 1,038,391, 1957.
12 A. Bokranz and H. Plum, Topics Curr. Chem., 16 (1971) 365-403.
13 U. Thust, Tin and its Uses, 122 (1979) 3-5.
14 H.G. Nösler, Gesundheitswesen u. Desinfektion, 62 (1970) 65-99; 175-176.
15 P.B. Hudson, G. Sanger and E.E. Sproul, Med. Ann. District Columbia, 28
    1959) 68.
16 G. Rees, S. African Med. J., 36 (1962) 3rd March.
17 G.A.J. Ayliffe, B.J. Collins and E.J.L. Lowbury, Brit. Med. J., 2 (1966)
    442-445.
18 H. Schulz-Utermöhl, Gesundheitswesen u. Desinfekt., 1967, 3rd April.
19 S.A. Edgar, Poultry Sci., 35 (1956) 64-73.
20 J.L. Fry and H.R. Wilson, Poultry Sci., 46 (1967) 319-322.
21 M. Graber and G. Gras, Rev. Elev. Med. Vet. Pays Trop., 18 (1965) 405-422.
22 J. Deufel, Fischwert, 20 (1970) 189-191.
23 C.A. Hall and P.D. Ludwig, Veterinary Rec., 90 (1972) 29-32.
24 W.P. Meleney and I.H. Roberts, Rec. Adv. Acarology, 2 (1979) 95-101.

Chapter 9

MONO-ORGANOTINS

9.1   INTRODUCTION

Mono-organotin compounds have one direct tin-carbon bond and have the general formula $RSnX_3$. This class of compounds has not yet found wide use in industry, despite the fact that the mono-organotins have a number of industrially interesting properties, not least of which is their low mammalian toxicity. LD_{50} values for some mono-organotin compounds are shown in Table 9.1.

TABLE 9.1

Acute oral toxicity values for some mono-organotins [1]

Compound	LD_{50} (rats) mg/kg
methyltin trichloride	575 - 1370
methyltin tri(isooctylthioglycolate)	920 - 1700
ethyltin trichloride	200 (LD_{100})
butyltin trichloride	2200
butanestannonic acid	> 4000
octyltin trichloride	2400 - 3800

Mono-organotin trihalides, from which other derivatives are made, are hygroscopic, low-melting solids or liquids which are to varying extents hydrolysed in water or moist air, liberating hydrogen halides. They are soluble in most organic solvents and in water containing enough acid to retard hydrolysis. These trihalides are strong Lewis acids and form complexes with ammonia, amines and many other oxygenated organic compounds; in many ways they resemble acid chlorides. The halogens are replaced readily by a wide variety of nucleophilic agents, hence the usefulness of these trihalides for preparing other derivatives. Monobutyltin oxide is a sesquioxide, $C_4H_9SnO_{1.5}$ from which it is difficult to remove the last traces of water. Butylstannonic acid is slightly acidic and forms alkali-metal salts, the alkylstannonates when excess alkali is used to hydrolyse the organotin trichloride. When organotin trihalides are treated with alkali metal sulphides the sesquisulphide forms; monobutyltin sesquisulphide is a tetramer, $(BuSnS_{1.5})_4$, and is commercially used as a PVC stabiliser.

9.1.1 Manufacture

Industrial processes for manufacture of mono-organotin compounds start from either tin(II) halides or from tin(IV) halides. Anhydrous tin(II) halides react with alkyl halides at elevated temperatures to form the mono-alkyltin trihalides:

$$RX + SnX_2 \longrightarrow RSnX_3$$

Good yields are obtained when a catalyst such as a trialkylantimony compound is present; better yields are obtained with bromides than with chlorides. A significant advance was the preparation of β- substituted ethyltin trihalides in good yields by reacting tin(II) chloride, hydrogen halides and α,β - unsaturated carbonyl compounds in common solvents at room temperature and atmospheric pressure [2].

If tin(IV) chloride is used as a starting material, the mono-alkyltin trichlorides are usually prepared by heating with the appropriate tetra-alkyltin compound in a 1 : 2 mole ratio at about 100°C.

$$R_4Sn + 2SnCl_2 \xrightarrow{100°C} 2RSnCl_2 + R_2SnCl_2$$

The products may be separated by vacuum distillation or may be utilised directly as a mixture to produce synergistic blends for PVC stabilisation.

9.2 APPLICATIONS

9.2.1 Stabilisation of polyvinyl chloride

Only one mono-organotin, monobutyltin sesquisulphide, has been used commercially in its own right as a stabiliser for polyvinyl chloride (PVC). This compound is approved for use in food contact PVC in a number of countries. It is also used in conjunction with dialkyltin thiotins in synergistic mixtures.

The use of mono-organotin compounds in synergistic mixtures with dialkyltin compounds for stabilising PVC is described in some detail in Chapter 2. This synergistic effect was first described in 1969 [3]. In particular the incorporation of about 30% of a mono-organotin in a thiotin mixture gives optimum early colour control during thermal processing of PVC.

In studies at the International Tin Research Institute [4] and at Akzo Chemie [5], ^{1}H and ^{119}Sn NMR spectroscopy have been used to follow reactions occurring between mono-alkyltin and dialkyltin species in thiotin stabiliser systems in solution, and infra red spectroscopy has been used to study solid PVC films

Fig. 9.1. Preparing PVC films to study the use of mono-organotins as stabiliser synergists.

(Fig. 9.1). The mono-alkyltin compound was shown to undergo a facile exchange of its mercaptide groups with chlorine atoms from the corresponding alkyltin chlorides, R_2SnCl_2 or $RSnCl_3$:

$$R_2SnCl_2 + RSn(IOTG)_3 \longrightarrow R_2Sn(IOTG)Cl + RSn(IOTG)_2Cl$$

$$RSnCl_3 + 2RSn(IOTG)_3 \longrightarrow 3RSn(IOTG)_2Cl$$

where IOTG is isooctylthioglycolate. In the case of estertin compounds, Dworkin [5] has presented evidence to show that with β- carboalkoxyethyltin compounds, the positions of the respective mercapto-ester/Cl exchange equilibria are dominated by carbonyl-to-tin coordination from the ester group of the β- carboalkoxyethyl moiety.

9.2.2 Treatments for glass surfaces

A new interesting use for monobutyltin trichloride is in the production of thin, transparent films of tin(IV) oxide on glass. The organotin is vaporised and brought into contact with the glass surface in the presence of air, at about 600°C, whereupon decomposition and oxidation to tin(IV) oxide occurs at the glass surface. Tin(IV) chloride is widely used in glass treatment processes, but the use of organotin compounds is claimed to eliminate problems such as fuming and corrosion and to give increased coating efficiency.

The development of tin chemical treatments for strengthening glass was reviewed by Evans [6]. The fact that thin films of tin(IV) oxide less than 100 nm
thick would increase the strength and impact resistance of glassware was known
as long ago as 1945 but it was not until the early 1960s that a process was developed commercially. Glass bottles are normally formed by blowing molten glass
to fill the contours of a mould, and the newly formed bottles, still hot, are
treated with a tin chemical which pyrolyses to the oxide at the glass temperature and forms a uniform thin film on the glass. Initial "hot end" processes
used tin(IV) chloride, either as a spray or in vapour form to produce the oxide
coating. The exact mechanism of the way in which the glass is strengthened is
still not fully understood. A diffusion process is considered to occur between
the compound and the glass so that some penetration of the glass surface occurs.
The treatment probably reduces the occurrence of minute surface cracks which appear as soon as the glass starts to cool and which have a weakening effect on
the structure. In addition, the tin oxide coating helps to bond an organic
lubricant, which is applied once the glass has cooled - a "cold end" treatment.
The introduction of wash-resistant cold end coatings in conjunction with the hot
end treatment in the early 1970s led to an extension of the process to multiple
use outlets, including milk bottles and catering tableware in addition to the
"one-trip" soft drinks bottles for which it was originally intended.

Although tin(IV) chloride is very effective in producing these coatings,
there have been reports of problems with corrosion of metal structures and
machinery by the hydrochloric acid formed during pyrolysis. The chloride also
reacts readily with moisture in the air and there is a tendency for coating
lines to clog. In view of this various manufacturers have looked at the possibilities of organotin compounds for producing oxide coatings. One Japanese company has developed a process using dimethyltin dichloride for glass treatment
[7]. Dimethyltin dichloride is a solid and it is first heated to vaporise it,
then mixed with air and blown onto the hot glass surface in the normal manner.
The use of a mixture of 20% monomethyltin trichloride with 80% dimethyltin dichloride has been described [8].

Recent work has shown that a liquid mono-alkyltin compound, monobutyltin
trichloride, may be pyrolysed on glass to produce tin(IV) oxide films. The process has been developed by the U.S. firm M and T Chemicals, Inc. [8] who also
evolved a new system for application of the chemical. The process, known as the
"Certincoat" * coating system, has now been commercialised and is being used
successfully by a number of manufacturers of glassware (Fig. 9.2).

*
    Trade name.

Fig. 9.2. Freshly made bottles entering the closed loop unit for deposition of a thin film of tin(IV) oxide.
Photograph courtesy: M and T Chemicals, Inc., Rahway, NJ, U.S.A.

The system includes a specially engineered metering pump and a double loop coating hood. The inner loop receives and recirculates the coating chemical vapour, whilst the outer loop acts as an air curtain and only circulates low concentrations of the coating chemical. Finish protection is ensured by directing a stream of fresh air at the top of the bottles. The monobutyltin compound has a low vapour pressure which makes handling easier and in the application system it is less susceptible to hydrolysis and fuming. Coating efficiency and productivity are claimed to be improved, by companies who are using the new process.

9.2.3 Homogeneous catalysts

Mono-organotin compounds have potential as homogeneous catalysts for esterification and transesterification reactions. At least one monobutyltin derivative is in commercial use for catalysing the production of dioctyl phthalate from iso-octanol and phthalic anhydride. There do not appear to be any problems with corrosion of steel reaction vessels, since the mono-organotin functions as a neutral catalyst. At the International Tin Research Institute, the transesterification reaction between butyl propionate and excess methanol, in the presence of various mono-alkyltin catalysts, has been studied, the reaction product being analysed by gas liquid chromatography. Typical results are shown in Table 9.2.

254

TABLE 9.2

Catalytic effect of mono-alkyltin compounds for:
BuO·CO·Et + MeOH $\rightleftharpoons$ MeO·CO·Et + BuOH

Catalyst 0.2M%	Conversion after 2 hour reflux %
No catalyst	0
$BuSnCl_3$	39
$OctSnCl_3$	28
$BuSn(OSiPh_3)_3$	0
$(Et_4N)_2^+ [BuSnCl_3Br_2]^{2-}$	37
$BuSn(OH)Cl_2 \cdot H_2O$	14

If the mono-alkyltin catalyst is incorporated into the butyl ester molecule, e.g. $BuO·CO·CH_2·CH_2SnCl_3$, the methanolysis reaction (to form $MeO·CO·CH_2·CH_2SnCl_3$) is complete within 2 hours.

9.2.4 Textile treatments

The effectiveness of certain mono-organotins as flame resist treatments for woollen fabrics has been studied at the International Tin Research Institute [9]. Treatments based on mono-organochloro- and mono-organobromo stannates conferred an adequate flame resistance to wool gaberdine fabrics at tin levels of less than 1%. Studies by ^{119m}Sn Mössbauer spectroscopy indicated that in the case of the compound $(Me_4N)_2^+ (BuSnCl_5)^{2-}$, the active species is a halogen-containing mono-organotin oligomer which is fixed in the fabric at 70°C. However, on washing, hydrolysis of the tin-halogen bonds occurs and the treatment becomes ineffective, illustrating that the presence of the halogen is necessary for flame resistance.

Certain long-chain mono-n-alkyltin compounds exert water repellent properties and they have been evaluated in cotton fabric using a standard vertical spray test (B.S. 3702 : 1982). Typical results are shown in Table 9.3 for de-sized and bleached cotton fabric.

Studies of the cloth by electron microscopy and ^{119m}Sn Mössbauer spectroscopy have indicated that in the case of treatment with aqueous sodium butanestannonate, the mono-alkyltin compound is hydrolysed in the fabric to produce an insoluble deposit of butanestannonic acid, $(BuSn(O)OH)_n$ [10].

TABLE 9.3

Water repellency of mono-organotin treatments on cotton fabric

Compound	Treatment conditions	B.S. 3702 : 1980 Spray rating *
Na$^+$ [BuSn(O)O]$^-$	0.1M in water	4
BuSn(OH)$_2$Cl	0.1M in methanol	4
BuSn(OSiPh$_3$)$_3$	0.1M in acetone	4
OctSn(OH)$_2$Cl	0.1M in acetone	4
Untreated	-	1

* 1 = Complete wetting of whole surface.
  5 = No wetting and no adherence of small drops.

Fig. 9.3. British Standard test for assessing water-proofing effectiveness, in use to assess a mono-organotin treatment on wool.

Plum [11] has also conducted water repellency studies on cotton canvas. Treatments based on 2% butyltin and octyltin trichlorides and triacetates proved as effective as a commercial silicone compound and were even better than the silicone at lower (0.5%) concentrations. Plum also studied the hydrophobic action of these mono-organotins on building materials such as concrete or limestone and on wood. Typical results for limestone are shown in Fig. 9.4.

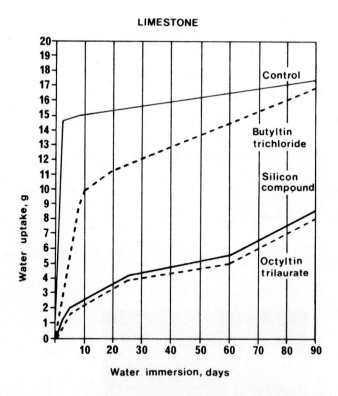

LIMESTONE

Fig. 9.4. Comparative results of water repellency tests for limestone [11].

A test has been devised at the International Tin Research Institute for assessing the effectiveness of waterproofing treatments in preventing rising damp in bricks. A stack of three bricks, interleaved with absorbent paper, is allowed to stand in half an inch (1.25 cm) of water; the middle brick is treated with the water repellent compound under test and its effectiveness as a barrier layer is monitored by periodic weighing of the topmost brick. Initial results indicate that the mono-alkyltin compounds have a significant water-repellent action, the longer chain mono-n-alkyltin compounds being effective at the lowest concentrations, e.g. 0.003M [12]. Studies have also been conducted on the possible use of aqueous solutions of sodium butanestannonate as flotation agents for Cassiterite and related ores, but results to date have been variable. Examination of the ^{119}Sn NMR spectrum of the monobutyltin compound in water indicates that the tin moiety is present 'as a hexacoordinate anion of the type $\left[BuSn(OH)_5\right]^{2-}$, rather than as the expected $\left[BuSn(O)O\right]^{2-}$ and this hydrophobic species seems to be rather readily adsorbed on certain ores.

9.2.5 Fungicidal properties

In general, mono-organotin compounds do not exhibit significant biocidal properties, in fact they are notably low in toxicity, particularly towards mammals. However in one patent [13] fungicidal activity is claimed for monobutyltin- and monophenyltin triformates, together with low mammalian toxicity (LD_{50} values (rats) of 400 and 350 mg/kg, respectively). In spore germination tests against Phytophthora infestans, Alternaria tenius and Botrytis cineria (typical organisms attacking crops), monophenyltin triformate gave results comparable with triphenyltin acetate, a commercial fungicide. Monobutyltin triformate had inhibiting effects on wood-rotting fungi even at concentrations of 0.01%. In studies at the Institute for Organic Chemistry TNO, Utrecht, monophenyltin stannatrane, $PhSn(OCH_2CH_2)_3N$ was shown to be active against a range of fungal organisms [14]. Organic silatranes are well known to have a wide spectrum of biological action, so that stannatranes might be expected to display a similar trend.

REFERENCES

1   L.A. Hobbs and P.J. Smith, Tin and its Uses, 131 (1982) 10-13.
2   W. Germain, H.P. Wilson and V.E. Archer, Encyclopaedia of Chemical Technology, 23 (1983) 3rd edn., Publ. by John Wiley and Sons, Inc., 42-77.
3   P. Klimsch and P. Kühnert, Plaste u. Kautschuk, 16 (1969) 242.
4   International Tin Research Council, Ann. Report 1980, 22-23.
5   J.W. Burley and R.E. Hutton, Polym. Degrad. Stabil., 3 (1980/81) 285.
6   R.D. Dworkin, Tin and its Uses, 139 (1984) 7.
7   C.J. Evans, Glass, Sept. 1974, 303-305.
8   T. Suzukawa, Ceramics, 4 (1969) 852-855.
9   Cincinnati Milacron Chemicals, "Methyltin chloride 8020", Tech. Bull., 1976.
10 U.S. Patents 4,130,673 : 1978 and 4,144,362 : 1979.
11 P.A. Cusack, L.A. Hobbs, P.J. Smith and J.S. Brooks, J. Text. Inst., 1980 (3) 138-146.
12 International Tin Research Council, Ann. Report 1982, 22.
13 H. Plum, Tin and its Uses, 127 (1981) 7-8.
14 Brit. Pat. 1,081,969 : 1967.

Chapter 10

ENVIRONMENTAL ASPECTS OF ORGANOTINS

10.1   INTRODUCTION

Tin is a non-toxic metal and indeed its major use, as a thin metallic coating
on steel sheet to make tinplate for canning applications, depends on its safety
in contact with food. Most inorganic tin compounds, too, are non-toxic. Yet
the formation of one or more tin-carbon bonds has a profound effect on the behaviour of the resulting organotin compounds. As has been shown in preceding
chapters, many organotin chemicals exert very powerful biocidal activity, whilst
others serve as stabilisers and catalysts in the plastics industry and possess
varying degrees of toxicity.

The influence of organotin compounds in the environment has to be carefully
considered in view of the increasing industrial use and the variety of applications of these chemicals. This applies especially to triorganotin compounds,
since these have the greatest effect on micro-organisms and on higher beings,
such as warm-blooded animals and fish. There is good evidence to suggest that
organotins ultimately degrade in the environment to produce harmless inorganic
forms of tin. In this respect they are unlike certain other heavy metals such
as lead and mercury whose use imposes a permanent polluting burden. However,
safe long-term use of organotin compounds requires a thorough understanding of
their toxicological properties and their behaviour in the environment.

The world's manufacturers of organotin compounds have taken a responsible
initiative, in that in 1978 they set up a body, the Organotin Environmental
Programme (ORTEP) Association whose aim is to promote and foster the dissemina-
tion of scientific and technical information on the environmental effects of organotin compounds. Members of this Association meet regularly and a Data Bank
on environmental aspects of organotins is maintained.

10.2   STRUCTURE/BIOCIDAL ACTIVITY RELATIONSHIPS

The toxicological properties of organotins are in general dependent upon the
number of organic groups introduced at the tin atom; thus in a series R_nSnX_{4-n},
maximum biological activity occurs when n = 3, i.e. in the triorganotins R_3SnX.
Within the R_3SnX series, whereas the nature of the group X does not usually have
much influence on the biocidal properties, especially when it is an inorganic
group, the organic group R does have an effect on the toxicity pattern. Thus

trimethyltin compounds show maximum activity against insects, triethyltin compounds have maximum toxicity for mammals and tributyltin compounds are most active against fungi, molluscs, gram-positive bacteria, fish and plants. The pattern is shown clearly for trialkyltin compounds in Fig. 10.1; a significant
feature is that the maximum distance between mammalian and fungicidal activity
occurs in the case of tributyltin compounds, which explains their widespread
adoption in wood preservatives.

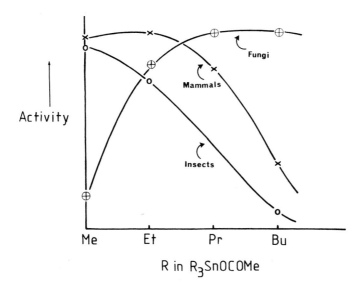

Fig. 10.1. Dependence of the biological activity of tri-n-alkyltin acetate on
the nature of the alkyl group for different species.

The underlying cause of the broad spectrum of activity shown by the triorganotin compounds is thought to be derangement of mitochondrial functions. The
mechanisms which may be involved have been summarised by Blunden et al. [1] as
follows:

(a)   Interaction with mitochondrial membranes to cause swelling and disruption.

(b)   Secondary effects due to their ability as ionophores to derange mitochon-
       drial function through mediation of Cl^-/OH^- exchange across the lipid mem-
       brane.

(c)   inhibition of the fundamental energy conservation processes involved in
       the synthesis of ATP from ADP, for example by mitochondrial oxidative
       phosphorylation and also photosynthetic phosphorylation in chloroplasts.

Triorganotin compounds can react with active centres in the cell by forming co-ordinating links with amino acids of the cell proteins. Smith [2] has discussed the relationship between the structure of organotin compounds and their biocidal activity. He summarised studies conducted to determine possible binding sites for triorganotin compounds on simple amino acids or dipeptides as an indication of the mode of attachment of triorganotins to protein molecules. He concluded that the most probable binding sites appear to be via HS— and amidazole NH— groups:

$$R_3SnX + \begin{cases} HS— \\ HN— \end{cases} \longrightarrow \begin{cases} R_3SnS— & + HX \\ R_3SnN— \end{cases}$$

An intramolecularly chelated tributyltin dialkylaminoalkoxide has been shown to be much less toxic to mice than bis(tributyltin) oxide and it is considered that the toxicity of the trialkyltin compounds R_3SnX may be independent of the X radical only when this is a simple non-chelating group which is capable of exchanging at these active protein sites. The di-n-alkyltin compounds R_2SnX_2 show a trend towards decreasing toxicity with increasing length of the alkyl chain and certain of the lower homologues are able to inhibit the oxidation of α-keto acids by combining with co-enzymes possessing vicinal dithiol groups. Their mode of action is quite different from that of the triorganotins and the anionic group X appears to influence the biological activity in the diorganotins. The dioctyltin compounds have a low toxicity as befits their use as stabilisers in food-contact PVC. All the mono-organotin compounds have low mammalian toxicity, as far as is known. In the case of the tetra-organotins, these have a delayed toxic action and there is some evidence to indicate that in mammals they may be converted to the triorganotin species, particularly in the liver [3]. Although used to some extent in antifouling paints, the major use of triphenyltin compounds is in agriculture, where they are active against a range of fungi and insect pests whilst having sufficiently low phytotoxicity to be used on crops. Extensive details of research on the toxicity, excretion and accumulation, and metabolism of these compounds are given in a review by Bock [4]. Mitochondria and erythrocytes, suspended in sodium chloride or ammonium chloride solution, swell after triphenyltin chloride or hydroxide has been added and the permeability of the mitochondrial membranes to Na^+, K^+ and Cl^- ions, as well as to fumarate, malate and citrate increases. Like the trialkyltin compounds, the triphenyltins inhibit oxidative phosphorylation of ADP to ATP in mitochondria and chloroplasts. Inhibition of the $Na^+ - K^+$ ATPase in cell membranes and of ATPase for Ca^{2+} transfer in sarcoplasmic reticulum has been found. In compounds of the type R_3SnX, the influence of the group X on toxicity is low, although this finding has been disputed by some workers. Systematic experiments with triaryltin compounds in which the aryl group was varied led to no improvement over the biocidal effectiveness of the triphenyltin compounds.

10.3   TOXICOLOGY OF ORGANOTIN COMPOUNDS

There is now an extensive literature on the toxicity of organotin compounds, but it is often difficult to compare the findings because of the great differences in the design of the experiments, for example the purity and form of dosage of the compound and the mode of administration (oral, percutaneous, intraperitoneal, intravenous) and the species used (cat, mouse, rat, etc.).   There are also differences in the number of administrations, the period of observation and the number of animals employed per experiment.   One of the most comprehensive reviews of toxicity and toxicity data has been compiled by Smith et al. [5] .

The basis for an initial toxicological assessment and classification is the determination of the acute oral LD_{50} value with a single administration of the test substance, usually to rats or mice.   (The LD_{50} value is the amount after whose administration about 50% of the test animals in a test set, die.)   Of equal value are investigations that determine the sub-acute, sub-chronic and chronic toxicity.   The latter provides information on cumulative effects, such as build-up in animal tissues.   Low dosages of the test substance are used to determine its fate in the animal organism, including its chemical transformation and excretion. LD_{50} values for some industrially used organotin compounds are listed in Table 10.1.

10.3.1   Tetra-alkyltin compounds

Although not biologically active, tetra-alkyltin compounds are toxic to mammals with a maximum activity occurring at tetra-ethyltin.   With increasing alkyl chain length, the toxicity diminishes.   The low tetra-alkyltin compounds would be classified as moderately toxic, but their high volatility and the absence of a characteristic odour in the case of the pure compounds makes them potentially dangerous.   Symptoms of poisoning are similar to those for the trialkyltin compounds and this fact, together with the delay in onset of symptoms, suggests a conversion of tetra-alkyltin compounds into trialkyltin compounds in the animal organism.

10.3.2   Trialkyltin compounds

The most toxic triorganotin compounds to mammals are the triethyltins; the principal site of action seems to be the central nervous system.   All toxic trialkyltin compounds except the methyl derivative produce an interstitial oedema in the white matter of the brain and spinal chord, without a histologically detectable change in the neuronal tissue.   The mechanism is thought to involve

TABLE 10.1

Acute oral LD_{50} values for some industrial organotin compounds [5]

Compound	LD_{50} * mg/kg	Test animal m = mouse r = rat
Tetrabutyltin	> 4000	r
Bis(tri-n-butyltin) oxide	148-234	r
Tri-n-butyltin fluoride	200	r
Tri-n-butyltin acetate	125-380.2	r
Tri-n-propyltin oxide	120	r
Triphenyltin fluoride	486	m
Triphenyltin chloride	118-135	r
Triphenyltin hydroxide	108-360	r
Triphenyltin acetate	125-491	r
Tricyclohexyltin hydroxide	235-540	r
Fenbutatin oxide	2630	r
1-tricyclohexylstannyl-1,2,4-triazole	631	r
Di-n-octyltin-S,S'-bis (isooctylthioglycolate)	1200-2100	r
Di-n-octyltin maleate	4500	r
Di-n-butyltin isooctylthioglycolate	500-1037	r
Di-n-butyltin oxide	487-520	r
Di-n-butyltin diacetate	109.7	m
Di-n-butyltin dilaurate	175-1600	r
Mono-octyltin trichloride	2400-3800	r
Mono-octyltin isooctylthioglycolate	3400->4000	r

* A range indicates highest and lowest value for LD_{50} reported in the literature for the compound.

membrane damage. Mild cases of poisoning in animals and in man, recover completely.

Trialkyltin compounds are used in wood preservatives, antifouling paints and as disinfectants. Any potential health hazards will relate to handling, and provided manufacturers' precautions are followed, there should be no risk involved. Contact with the skin and inhalation of vapours should be avoided and good ventilation and the wearing of safety gloves and goggles is necessary. Contaminated working clothes should be changed at once; hands should be thoroughly cleansed before any interruption to work occurs (for meals, smoking, etc.).

10.3.3   Triphenyltin compounds

Since these chemicals are widely used in agricultural applications, it follows that their mammalian toxicity must be low. There have been isolated instances of skin irritation and other reactions when the chemicals have been incorrectly handled. The most serious accidents reported to date resulted when a combined preparation of triphenyltin acetate and maneb, being sprayed from an aeroplane, was incorrectly handled. Two pilots exhibited symptoms of poisoning and three mechanics suffered slight ill effects. The poisoning was character- ised by an inflammation of the gastro-intestinal tract and disease of the paren- chyma, particularly in the liver. Symptoms disappeared after exposure ceased, although taking several months in one case.

10.3.4 Dialkyltin compounds

The symptoms of poisoning produced by dialkyltin compounds are completely different from those which result from trialkyltin compounds. In rat studies at moderate oral doses, the dibutyltin compound was the most toxic, whereas at higher doses, the methyl, ethyl and propyltin compounds also caused the death of rats. Dioctyltin dichlorate, which had been as toxic as dibutyltin dichloride when given intravenously, showed practically no toxicity in oral administration. Contact with the skin produced lesions in the case of all compounds up to and including dibutyltin. Dialkyltin dichlorides at toxic doses produce an unchar- acteristic general illness in rats, in particular an inflammatory lesion of the bile duct. Dialkyltin compounds with alkyl groups longer than butyl have low enough toxicity to be used as stabilisers in PVC for food packaging. Dimethyl- tin compounds have recently been proposed for the stabilisation of PVC, the LD_{50} of dimethyltin-S,S'-bis(isooctylthioglycolate) falling within the range of LD_{50} for dioctyltin compounds. The purity and stability of the stabiliser is of par- ticular importance, in order to avoid the presence of toxic trimethyltin com- pounds.

10.3.5 Metabolic fate of organotin compounds

Blunden [1] has reviewed recent work on the metabolism of organotin compounds in mammals. Studies on diethyltin species showed that they are broken down in vivo in rats to mono-ethyltin salts which are rapidly eliminated from the body. Studies in rats and mice using tricyclohexyltin, triethyltin and tributyltin compounds have indicated that they are excreted essentially intact from animals. However, another study has indicated that tricyclohexyltin hydroxide is meta- bolised in vivo in rats to di- and mono-organotin and inorganic tin species [6]. In vitro studies using rat liver microsomal mono-oxygenase enzyme systems have indicated that the primary metabolic reaction for butyltin compounds is carbon hydroxylation of the butyl groups where the α- and β- carbon-hydrogen bonds are found to be more susceptible to hydroxylation.

Bock [4] has summarised studies on the metabolism of triphenyltin compounds in mammals. The degradation of triphenyltin chloride with time, in the organs of rats fed a single oral dose of 3 mg, was followed by quantitative analysis of the degradation products in faeces and urine. There was a gradual increase in the concentration of di- and mono-phenyltin compounds at the expense of triphenyltin. This gradual breakdown was confirmed by other workers. There have been several reports of the formation of diphenyltin compounds on plants which have been treated with triphenyltins; since benzene has been found as a breakdown component of triphenyltin compounds, it can be concluded that degradation in warm-blooded animals as well as in plants, proceeds along the path:

$$(C_6H_5)_3Sn^+ \longrightarrow (C_6H_5)_2Sn^{2+} \longrightarrow (C_6H_5)Sn^{3+} \longrightarrow Sn^{4+}$$

The general trend indicated in a large number of metabolic studies, is that organotin compounds are not cumulative, being readily excreted even when large doses are administered. Non-excreted compounds probably degrade down to inorganic tin, leading to a progressive detoxification.

10.4   EFFECTS OF ORGANOTIN IN THE ENVIRONMENT

The principal possible sources of entry of organotins into the environment are shown in Fig. 10.2. Each of these will be examined in turn and the potential ecological burden assessed in the light of current practices and the known behaviour of organotin compounds. However, it will be useful first to discuss some properties of organotins which are relevant to the issue.

10.4.1   Stability of organotin compounds

Whereas in R_3SnX or R_2SnX_2 compounds, the anionic group X can be easily separated through hydrolysis, the tin-carbon bond is considerably more stable in alkyltin compounds and can generally only be broken by drastic treatments such as strong acids or alkalis. However, the triaryltin compounds such as triphenyltin chloride or fluoride are even sensitive to weak acids. Oxidising agents such as potassium permanganate are also capable of breaking down organotin compounds and this has been used to inactivate wastes containing the compounds. The process consists of a progressive separation of organic groups leading eventually to inorganic tin [6]:

$$R_4Sn \longrightarrow R_3Sn- \longrightarrow R_2Sn< \longrightarrow RSn< \longrightarrow SnO_2$$

The industrially used organotin compounds show considerable thermal stability, the organic moiety splitting off at temperatures between 180 and 200°C for tetra, tri- and di- compounds and at 120°C for monobutyltin chloride. At 280°C T.B.T.O. breaks down almost completely to inorganic tin after 5 hours. When wood containing T.B.T.O. is burned, no tributyltin can be detected in the residue, only inorganic tin and small amounts of di- and mono-organotins.

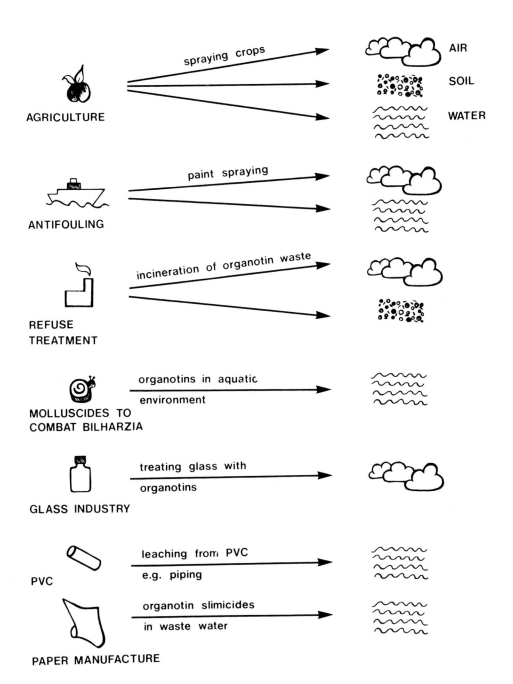

Fig. 10.2. Possible sources for introduction of organotins into the environment.

Organotin compounds are decomposed by light to varying degrees, depending partly on the wavelength of the light with shorter wavelengths having most effect. Under the influence of ultra-violet light, degradation is accelerated and a large number of studies have been conducted with organotins in soil, in the pure form and in aqueous solutions (restricted by the insolubility of most organotin compounds). The results demonstrate a progressive degradation of triorganotins to di-, mono-, and inorganic species. Ultra-violet breakdown is one of the most significant modes of degradation of organotins in the environment. Blunden [6] has studied the ultra-violet degradation of methyltin chlorides in carbon tetrachloride and in water and has established approximate relative rates of degradation for the various species. The final product was shown to be hydrated tin(IV) oxide.

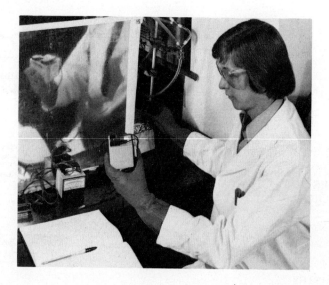

Fig. 10.3. Equipment for studying the breakdown of organotin compounds under irradiation by ultra-violet light, at the International Tin Research Institute.

Certain fungi and bacteria are able to break down organotin compounds, particularly tributyltin and triphenyltin compounds. These include Pseudomonas aeruginosa bacteria which can decompose T.B.T.O. under aerobic conditions and the wood-rotting fungus Coniophora puteana which is also capable of attacking tin-carbon bonds.

10.4.2 Sources of entry of organotins into the environment

The principal uses of organotins have been described in the preceding chapters and comprise stabilisers for PVC (mono- and di-organotins), catalysts (diorganotins), and biocidal applications (triorganotins). It is only the latter group of uses which are likely to contribute significantly to the amount of organotins in the environment. Leaching studies on PVC have shown that the leaching rates for organotin stabilisers are very low. As regards waste disposal, incineration decomposes the organotins whilst the inclusion of PVC amongst domestic refuse for landfill operations would involve very low concentrations indeed of the organotin. Again, the level of organotin in flexible polyurethane foam products is only of the order of 0.02%, so that this outlet is not a serious source of pollution.

One area where organotins are actively introduced into the environment is in agriculture, where crops are sprayed to protect them against fungal or insect infestation, sometimes from aircraft. The possibilities of an agricultural chemical entering the food chain at a toxic level are carefully examined before a new product is permitted to be used (see Chapter 6). This involves determination of pesticide residues on plants and in animal feeds, levels found in milk and other animal products, etc. The concentration of triphenyltin compounds on treated plants decreases rapidly because of losses in wind and rain and degradation in the atmosphere and under the influence of light. Field experiments have shown that the organotin level is usually reduced to a half within 3-4 days. The residues remaining on plants which are eaten by humans do not normally constitute any risk, since the parts of the plants used for human foods (potatoes, carrots, celery, rice, etc.) do not normally contain detectable levels; moreover food processing generally completely or partially decomposes any residues. Tests on animal products from beasts fed treated sugar beet leaves showed levels of phenyltin compounds close to the detectable limit. It was concluded that human consumption of organotin compounds in meat was only possible when animals were slaughtered immediately after consumption of treated leaves; even then, levels of organotin would be very low and cooking would reduce them further. Waiting times are in any case specified after spraying of crops, before harvesting can begin, and permissible limits of residues are strictly fixed.

Triorganotins can reach soil and water via agricultural application, via antifouling paints and from wood treated with organotin preservatives (although the biocide is not easily removed from the wood). Organotins in contact with soil are fairly rapidly decomposed, mainly by micro-organisms in the soil and by oxygen in the air. The compounds have been shown to be adsorbed readily on the soil and are not easily leached away. The effect of T.B.T.O. (10 and 100 ppm) on the microbial activity relevant to fertility of soil has been studied [7]. The influence on soil microbial populations, ammonification, nitrification, sulphur oxidation and soil respiration was studied and it was concluded that levels of T.B.T.O. up to 100 ppm had no biologically significant impact on soil micro-organisms or on their influence on soil fertility.

Release of toxicant from antifouling paints is a source of organotins in water, primarily sea-water. The main contamination occurs when a coated vessel is stationary, for example in harbours and bays. At the Fourth International Conference on the Organometallic and Co-ordination Chemistry of Germanium, Tin and Lead, held in Montreal, Canada, in August, 1983, a number of papers were given which were concerned with the fate of organotin compounds in the aqueous environment. These have been summarised [8]. Sherman studied the degradation of tributyltin fluoride and chloride at 200 ppb in aqueous media (distilled water and various buffer solutions). He concluded that degradation progressed through hydrolysis of the halide bond to form a tributyltin hydroxy species which then dehydrates to tributyltin oxide and subsequently degrades to dibutyltin oxide. The halides and the degradation products were estimated to have a life of about 60 days in aqueous media. Maguire studied the fate of tributyltin oxide in the aquatic environment in the light of reports that butyltin species have been found in the waters and sediments of some harbours in Ontario. The tributyltin is considered to undergo slow photolytic decomposition in sunlight, at least partially by stepwise de-butylation to inorganic tin. Chau reported the presence of methyltin compounds in water samples from various lakes, rivers and harbours in Ontario. $MeSn^{3+}$ and Me_2Sn^{2+} were always detected, but Me_3Sn^+ was rarely found. Thayer studied the effects of alkyltin compounds on the growth curves of micro-organism populations from sediments. Dimethyltin dichloride did not affect the exponential growth phase, but did increase the lag time.

Although few systematic investigations have been carried out on the toxicity of organotin compounds in cold-blooded animals, it is known that fish are more susceptible to organotins than are warm-blooded mammals. Thus the lethal concentration (LC_{50}/48 hours) of T.B.T.O. for various species of fish was only 0.05 ppm. This fact should be borne in mind when using organotins. No serious effects have been reported, however, probably because of rapid detoxification in water, dilution effects, decomposition and adsorption on sediments. Further information on toxicity to fish is given in Chapter 5.

Sheldon [9] has proposed a generalised degradation scheme (Fig. 10.4) which covers tributyltin and triphenyltin compounds in soil/water environments and which probably holds also for other organotin compounds.

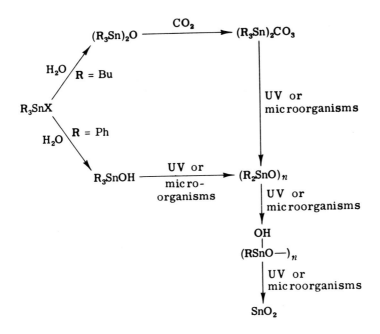

Fig. 10.4. Environmental degradation scheme for tributyltin and triphenyltin compounds [9].

10.4.3   Biomethylation of organotins

It is known that inorganic heavy metal compounds can be methylated under the influence of certain bacteria such as the _Pseudomonas_ series. Mercuric salts for example, can be converted into highly poisonous and volatile methylmercury compounds. Laboratory studies [10] have shown that certain tin(II) salts can be methylated by methyl-cobalamin under very specialised conditions (aqueous solution at pH 1 in the presence of Co^{3+} ions) to produce monomethyltin(IV) species. Smith [11] has pointed out that no reaction has been observed between tin(II) and methyl-cobalamin in the absence of an oxidising agent or between tin(IV) and methyl-cobalamin under a variety of conditions. The $SnCl_3^-$ species may be first oxidised to a trichlorostannyl radical $\dot{S}nCl_3$ which could then cleave the Co—C bond homolytically to produce $MeSnCl_3$. With advances in analytical techniques (see Table 10.2), a number of workers in the U.S.A. have been able to report the presence of methyltin species in natural media at very low concentrations.

Porter [13] has described work being conducted at the U.S. Naval Ship R. and D. Center. Strains of _Pseudomonas_ bacteria are being grown on agar surfaces treated with various organotin compounds, to see whether the bacteria can transform the tin compounds. The atmosphere above the bacteria is analysed by gas

TABLE 10.2

Techniques for determining inorganic and organotin compounds in waterways [12]

Compound investigated	Method	Approximate detection limit	Reference
R_3SnX (R = Me, Et and Bu) R_2SnX_2 (R = Me, Et and Bu) $RSnX_3$ (R = Me, Bu and Ph)	Conversion to hydride with $NaBH_4$, collection in a trap, separation by boiling-point, detection by AAS. (Atomic absorption spectroscopy)	1 ng/litre	V. F. Hodge, et al. Anal. Chem., 51 (1979) 1256.
R_3SnX (R = Me, Et, Pr, Bu, Ph and cyclo-C_6H_{11})	HPLC (high-performance liquid chromatography) strong cation exchange columns with determination by GFAA. (Graphite furnace atomic absorption spectroscopy)	5-30 ng as Sn	K. L. Jewett, et al. J. Chromatogr. Sci., 19 (1981) 583.
Me_nSnH_{4-n} (n = 1-4)	Automatic purge trap sampler followed by GC (gas chromatography) separation and determination by FID. (Flame ionization detector)	1 µg/litre on a 10 ml sample	J. A. Jackson, et al. Environ. Sci. Technol., 16 (1982) 110.
Me_nSnX_{4-n} (n = 0-3)	Conversion to hydride with $NaBH_4$, separation by boiling-point and determination by FED. (Flame emission detection)	<1 ng/litre on a 100 ml sample	R. S. Braman, et al. Anal. Chem., 51 (1979) 12.
Me_nSnX_{4-n} (n = 0-3)	Extraction with benzene, followed by butylation and separation by GC and determination by AAS.	0.04 ng/litre on a 5 litre sample	Y. K. Chau, et al. Anal. Chem., 54 (1982) 246.

Compound investigated	Method	Approximate detection limit	Reference
$Bu_n SnX_{4-n}$ (n = 1-3)	Solvent extraction, followed by methylation and separation by GC and determination by mass spectrometry.	10 μg/litre on a 1 litre sample	H. A. Meinema, et al. Environ. Sci. Technol., 12 (1978) 288.
$Bu_n SnX_{4-n}$ (n = 1-3)	Solvent extraction, followed by pentylation and determination with a modified FPD. (Flame photometric detector)	10 mg/litre on a 25 ml sample	R. J. Maguire, et al. J. Chromatogr., 209 (1981) 458.
$Ph_n SnX_{4-n}$ (n = 0-3)	Separation by solvent extraction of the individual species followed by determination as inorganic tin by photometric methods.	2 μg as Sn	K. D. Freitag, et al. Z. Anal. Chem., 270 (1974) 337.
$Ph_n SnX_{4-n}$ (n = 1-3)	Solvent extraction, followed by conversion to hydride with $LiAlH_4$ and separation and determination by GC - EC. (Gas chromatograph - electron capture detector)	0.015 μg/litre on a 200 ml sample	C. J. Soderquist, et al. Anal. Chem., 50 (1978) 1435.
$Bu_3 SnX$	Solvent extraction followed by direct photometric determination with dithizone.	1 ppm	L. Chromy, et al. J. Oil Col. Chem. Assoc., 51 (1968) 494.
$(Bu_3Sn)_2 O$	Solvent extraction and direct photometric determination on the haematein complex.	0.1 ppm	R. W. Pettis, et al. Austral. Def. Sci. Ser., Rept. No. 516 (1972).

Compound investigated	Method	Approximate detection limit	Reference
"Organotins"	Solvent extraction, wet-ash followed by photometric determination with phenylfluorone.	1 μg/litre on a 200 ml sample	L. R. Sherman, et al. J. Anal. Toxicol., 4 (1980) 31.
"Organotins"	Solvent extraction, wet-ash followed by photometric determination with phenylfluorone.	0.01 mg/litre on a 50 ml sample	E. Mor, et al. Annali de Chimica, 63 (1973) 173.
Bu_3SnX	Initial separation by steam distillation followed by extraction, wet-ashing and determination by inverse voltammetry.	0.01 ppm	H. Woggon, et al. Die Nahrung, 19 (1975) 271.
Ph_3SnX	Concentration by evaporation, extraction and determination by inverse voltammetry as above.	0.01 ppm	as above
Ph_3SnX	Solvent extraction and direct spectrofluorimetric determination with 3-hydroxyflavone.	0.004 ppm on a 50 ml sample	S. J. Blunden, et al. Analyst, 103 (1978) 1266.
Inorganic Sn(II) and Sn(IV)	Concentration by co-precipitation and flotation. Dissolved precipitate reacted with $NaBH_4$ and determination by AAS.	0.07 μg/litre on a 1 litre sample	S. Nakashima, et al. Bull. Chem. Soc. Jap., 52 (1979) 1844.
Inorganic Sn(IV)	Differential-pulse and anodic stripping voltammetry direct on the sample.	4 μg/litre on a 25 ml sample	G. Macchi, et al. Environ. Sci. Technol, 14 (1980) 815.

Compound investigated	Method	Approximate detection limit	Reference
Inorganic Sn(IV)	AAS/ETA direct on the sample but with the addition of 10% ascorbic acid before the ETA (electrothermal atomization) drying stage.	50 μg/litre	M. Tominaga, et al. Anal. Chim. Acta., 110 (1979) 55.

chromatography and other specialised techniques, to determine whether the bacteria are metabolising the tin and exhaling it in some form.

Blunden et al. [1] conclude that, whereas both tin(II) and tin(IV) compounds appear to be biologically methylated under simulated environmental conditions in sediments, chemical methylation may only be important for tin(II) species.

REFERENCES

1   S.J. Blunden, L.A. Hobbs and P.J. Smith, SPR 35 (Environmental Chemistry),
     Vol. 3, Chapter 2: Environmental Chemistry of Organotin Compounds, 1984,
     pp. 49-77.
2   P.J. Smith, Int. Tin Res. Inst. Publ. 596, 1978, 14 pp.
3   E.C. Kimmel, R.H. Fish and J.E. Casida, J. Agric. Food Chem., 25 (1977) 1.
4   R. Bock, Residue Reviews, 79 (1981), F.A. Gunther (Ed.), Publ. by Springer-
     Verlag, 1981, 270 pp.
5   P.J. Smith, J.G.A. Luijten and O.R. Klimmer, Int. Tin Res. Inst. Publ. 538,
     1978, 20 pp.
6   S.J. Blunden, J. Organometallic Chem., 248 (1983) 149-60.
7   W.B. Bollen and C.M. Tu, Tin and its Uses, 94 (1972) 13-15.
8   Tin and its Uses, 139 (1984) 6-8.
9   A.W. Sheldon, J. Paint Technol., 47 (1975) 54.
10   W.P. Ridley, L.J. Dizikes and J.M. Wood, Science, 197 (1977) 329.
11   P.J. Smith, Int. Tin Res. Inst. Publ. 569, 1980, 16 pp.
12   A.H. Chapman, Proc. Anal. Chem., 20 (1983) 210.

INDEX OF TIN COMPOUNDS

TRADE NAMES

The trade names listed here refer to organotin compounds
or active formulations containing organotins that are
mentioned in the text.